The Mirror Strategy

The Quiet Competitive Advantage of Digital Twins That Make AI Pay Off

BY

SAM SCHREIM

Important

Before Getting Started, Get the Most Out of This Book: Download the Visual Companion Pack

Bring this book to life with downloadable infographics, tables and slides. I've received feedback that the static visuals don't allow you to fully experience the strategies. That's why I'm providing a companion download, containing **dozens of visuals** of high resolution, easy-to-read templates, canvases, forms, checklists, cheat sheets and much more.

With these digital assets in hand, you'll be able to follow along and grasp each concept.

Unlock the full potential within these pages. Visit

www.dtaas.us/visuals

to download the complete visual companion set and actively apply what you learn. With crystal clear visuals at your fingertips, abstract ideas will spring to life, propelling your Digital Twin and ML skills.

Claim your visual package now – let's bring this book into full view together!

A CEO's Note: Read This First

AI seems to have developed a talent for presenting itself as brilliant when displayed on screens and questionable when viewed through the lens of a P&L statement. For instance, during an executive review of an operation, a large wall of indicators was glowing bright green: On-time delivery, up-time, quality, cost. Just minutes later, the operator called in to report that a single constraint had failed; the day's plan was now out-of-date. The data display had not been wrong; it had simply been designed to provide information on everything except the most important thing in that moment: What should we do now?

This awkward disconnect between AI performance (or lack thereof) and real-world results does not necessarily mean that AI is flawed; it typically means that the organization created the wrong type of AI-based system. Instrumentation and observability are critical components of any well-designed twin. The system must be measurable and monitored, though the interface need not be a traditional dashboard. Some twins operate through APIs, embedded control loops, or automated alerts with no human-facing display at all. What matters is that the system's state, predictions, and actions can be observed, traced, and audited. However, a dashboard is nothing more than an instrument panel; it is not a steering wheel. To turn visibility into actionable decision-making, organizations must find a way to govern their responses to the information they have gathered or remain as informed but still reactive as before.

The real risk

The real risk is not that the organization "misses AI." The real risk is that it buys AI the way it bought analytics: as reporting. That produces what this book calls Analytics Theater: The appearance of intelligence without the discipline of action.

The Mirror Strategy in One Sentence

A Digital Twin is not a prettier dashboard and not a 3D model built for a keynote. In this book, a Digital Twin is a living operational model, kept in sync with reality, that converts that synchronization into executable decisions.

A useful analogy is navigation. A paper map informs. A GPS guides. The map can be beautifully designed and still be wrong at the moment it matters. A twin, at its best, behaves like GPS: it tracks the current state, anticipates what is likely to happen next, and recommends a route that respects constraints.

Why this is a CEO's Advantage

A CEO does not need to out-code the engineers. In fact, the lack of

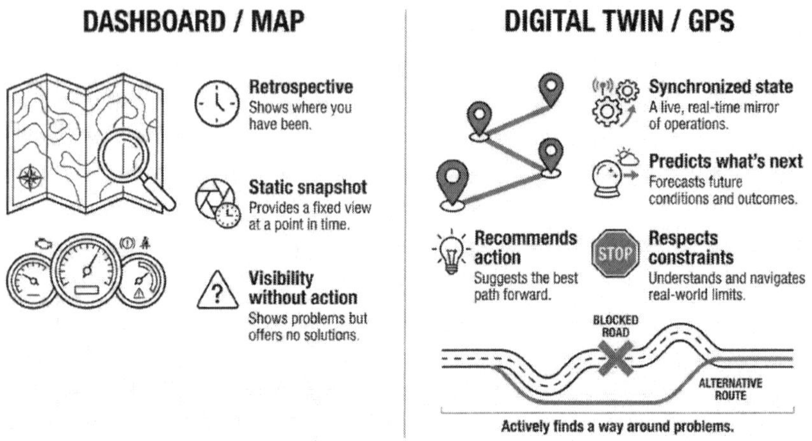

technical obsession is often an advantage. It forces the executive team to ask the questions that determine whether AI becomes value or vanity: What decision is being improved? What constraint cannot be violated? What outcome matters enough to measure? Who owns the decision when the model is right? Who owns the decision when it is wrong?

There is another question that matters just as much: Who should be making this decision in the first place? One of the least discussed advantages of a well-designed twin is its ability to push decision authority downward in the organization. When frontline workers can test decisions against a simulation before executing them, they no longer need to escalate routine choices to supervisors who are already overwhelmed. The twin validates what experience once had to approve. This shift does more than improve speed; it fundamentally changes how executives spend their time. Instead of approving daily operational decisions, they can focus on the strategic questions that actually require their judgment: where to expand, what to build, how to compete.

Four questions that keep AI honest

- **WHY** is this being done? The initiative should create competitive advantage: Cost, speed, reliability, differentiation, not just a faster demo.
- **WHAT** is the business value? Anchor it to a P&L measure: margin, cycle time, throughput, downtime, churn. State "how much, by when."
- **WHO** must be involved? The work is cross-functional by nature. Operations, data, product, finance, and risk must share one plan and one accountable owner.
- **WHEN** will it be proven or disproven? Time-box the first test. Set explicit kill rules. Scale what works. Retire what does not.

How to use this book

The chapters ahead are designed as a field manual, not a textbook. Readers stuck in dashboard mode can start with the Dashboard Trap and the AI Maturity Ladder. Readers curious about how twins change organizational dynamics, particularly how they free executives from daily firefighting, should read the Delegation Dividend chapter before diving into architecture. Readers ready to act can go straight to the Minimum Viable Twin chapter and work backward as needed. Readers worried about autonomy and risk can begin with governance: physics-informed guardrails, autonomy charters, and auditability.

The intent is simple: replace AI anxiety with a decision system. When the organization can prove one improved decision in a short cycle, confidence replaces hype. From there, the work compounds.

About the Author

Sam Schreim is a strategy obsessive first and a technologist second. For more than two decades, he has helped Fortune□500 CEOs, fast-moving startups, and government leaders tackle the same fundamental questions: *Where should we play? How do we win?* As AI moved from the research lab to the executive agenda, Sam wove machine-learning tools into his strategy playbook–turning data models into levers for margin, speed, and public-sector impact. Today his work sits at the intersection of classic competitive thinking and the new, AI-powered engines that now drive it.

Table of Contents

FOREWORD: THE QUIET ADVANTAGE....................1

INTRODUCTION: THE SILENT REVOLUTION....................5

The Real Issue Is Not the Model....................6
A Different Type of Mirror....................7
The First Twin Was Manual....................7
The Timeline of the Name....................8
From Theater to Operations....................10
The Quiet Revolution....................11

CHAPTER 1 – THE DASHBOARD TRAP....................13

Two Layers, Two Worlds....................15
The Latency Tax....................17

CHAPTER 2 – DEFINING THE MIRROR....................21

Navigation Analogy....................23
What a Digital Twin Isn't....................23
Temporal Connectivity....................24

CHAPTER 3 – THE AI MATURITY LADDER....................27

Level 1: The Reflective – What happened....................27
Level 2: The Diagnostic – Why did it happen?....................29
Level 3: The Predictive – What Will Happen Next?....................29
Level 4: The Prescriptive – What Should Be Done?....................30
Level 5: The Semi-Autonomous – Can We Execute Within Guardrails?....................30
How Twins Help the Climb....................31

CHAPTER 4 – THE DELEGATION DIVIDEND....................33

The Hierarchy of Authority....................34
The Flight Simulator Effect....................35
Proof in Practice....................36
The Executive Dividend....................37
The Immune System....................38

CHAPTER 5 – THE TWIN STACK....................41

Architecture of Value....................41
Layer 1 – The Boundary....................41
Layer 2 – The Truth Feed....................42
Layer 3 – The Model....................43
Layer 4 – The Decision Engine....................44
Layer 5 – The Action Loop....................44

CHAPTER 6 – IDENTIFYING THE OPPORTUNITY....................47

The Three Conditions Litmus Test 47
Where Twins Don't Belong 49
The Power Pair 49
The Cost Out and Productivity Twin (CPT) *51*
The Growth and Revenue Twin (GRT) *52*

CHAPTER 7 – PROOF IN THE NUMBERS 59

The Holistic Value Play 59
The Closed Loop 60
Unlocking Latent Capacity 61
Detecting the Hidden 62
Calculating The Business Case 63

CHAPTER 8 – THE MINIMUM VIABLE TWIN (MVT) 65

The 30-60 Day Strategy Playbook 65
Learning in the Shadows 67
An Option-Based Approach 69

CHAPTER 9 – THE DATA FOUNDATION 71

The Many Plumbing Problems 71
The Allure of the Platform 73
The Importance of Open Standards 74
Data Challenges 76

CHAPTER 10 – GOVERNANCE FOR SEMI-AUTONOMY 79

The Guardrail 79
The Autonomy Charter 83
The Explainability Requirement 85

CHAPTER 11 – THE CEO'S TOOLKIT AND THE FUTURE OUTLOOK 89

The Readiness Scorecard 89
Data Readiness *89*
Operational Clarity *91*
Cultural Receptiveness *91*
The Opportunity Canvas 91
A World of Controlled Semi-Autonomy 92

CONCLUSION: THE QUIET ADVANTAGE 95

REFERENCES 99

Foreword: The Quiet Advantage

"Most AI failures are quiet: nothing breaks, nothing changes."

I was five minutes into a site visit at a distribution warehouse just outside Boston when the brand-new autonomous forklifts threw their tantrum. A lone mis-scanned pallet sent the fleet into a frantic stop-start loop, wheels screeching until the air stank of burned rubber. The machines did exactly what they were trained to do. The operation did exactly what it could not afford: it froze.

This was not a failure of vision or of desire. This was a failure of design. The warehouse had purchased intelligent technology (the autonomous forklifts), but it had not acquired any means to control the unpredictable nature of reality (bad scans, delayed shipments, hurried workers). There was no way to elegantly absorb the messiness of reality and make it into a controlled decision.

Incidents such as this justify executives' debate about Artificial Intelligence (AI). Organizations will fall behind if they don't act and will be at risk of making costly mistakes if they rush into action with poorly designed solutions. Pilot projects are delivered by teams that have produced impressive results and subsequently stalled. Dashboards continue to grow. Chief Financial Officers (CFOs) begin asking where the value is being generated. The room goes silent. The end result is a modern paradox facing executives today: a sense of urgency without clarity.

The Quiet Middle Path and Bridging the C-Suite Paradox

The escape route is not more hype and not more hesitation. It is discipline. Not discipline as bureaucracy, but discipline as architecture: systems designed to move from insight to action, repeatedly, safely, and measurably.

The core mechanism is the Digital Twin: not a cinematic replica, but a living operational model that stays synchronized with reality and translates that synchronization into executable decisions. A dashboard reports what happened. A twin guides what to do next by testing options against constraints before the cost of being wrong becomes real.

A dashboard is a rear-view mirror. It tells an organization what happened. A twin is closer to a GPS with simulation capability: it shows current position, predicts what lies ahead, recommends a route, and allows the organization to test alternative paths before committing.

But perhaps the most underappreciated advantage of the twin is what it does to organizational structure. When a simulation can validate whether an action is safe, the person closest to the problem no longer needs to wait for approval from someone higher in the chain. The maintenance technician who notices a vibration anomaly can consult the twin, receive a validated recommendation, and act. The charge nurse managing patient flow can make bed allocation decisions within guardrails that the twin enforces. Decision authority shifts from hierarchy to verification. Executives who once spent their days approving routine operational choices find themselves with something rare: **time** to think about strategy.

Marvin Minsky's Pet Warning Still Haunts Us

AI pioneer Marvin Minsky made a joke about AI control back in 1970 when he said, "If we're lucky, they may want to keep us as pets." That joke has endured; however, its longevity does more than simply create humor; it creates an actual fear of being controlled through AI. The fear is not AI with a voice. It's AI that acts, especially in physical operations where errors are expensive and often irreversible.

To avoid slowing innovation down while avoiding handing over complete control to a black box, you need to develop control systems that are explicitly available to humans. This includes developing models of physics and policy; defining and establishing guardrails to define authority and establish who controls the system at any given time; and providing transparency and auditability to provide accountability (keeping track). An identical (or twin) system will serve this purpose well, because it allows the interrogation, correction, and improvement of the original system in the open.

The Pragmatic Middle the Headlines Miss

Much of today's discussion on AI is centered around language models and chat interfaces. These are both impressive and can be quite useful; however, these models do very little to affect the operational layers of an organization, where the vast majority of the value is either created or destroyed. To create operational advantage, you need to shorten the distance between signal and response: reduce the latency tax that converts minor issues into costly surprises.

Digital Twins are built to operate within the layers of an organization. They take a myriad of disparate signals and convert them into a singular representation of how work actually flows, apply analytics, optimization techniques, and if necessary, machine learning techniques to determine what should happen next.

What You'll Get

The Mirror Strategy is a practical playbook for leaders who want AI to show up in the income statement, fast. I begin where most programs fail: the Dashboard Trap. I then define the twin precisely, separating it from 3D theater, chatbots, and one-off simulations.

I then introduce an AI Maturity Ladder that clarifies the jump from reporting to diagnosis to prediction to prescription, and finally to semi-autonomous execution within guardrails. Before examining the technical architecture, I explore what may be the twin's most strategic benefit: its ability to push decision authority to the lowest competent level of the organization. This Delegation Dividend transforms how executives spend their time and how frontline workers experience their roles. I then break down the Twin Stack so teams can see what must be built, in what order, and why skipping the 'boring' layers (data plumbing, integration, semantics) makes the 'smart' layers confidently wrong.

I'll show how to choose the right starting point (i.e., decisions that repeat, live inside fixed constraints, and carry a high cost of variance) and then how to prove value with a Minimum Viable Twin in 30 to 60 days. The point is not to build a monument. The point is to buy an option: a small bet with a clear test, explicit kill rules, and a rapid path to scale for the winners.

Finally, I address the question that arrives right on schedule in every executive meeting: what prevents the system from becoming dangerously confident? The answer is governance: physics-informed guardrails, an autonomy charter that defines what the system may execute, and auditability so decisions can be explained, traced, and improved.

Competitors will not win by having better slogans. They will win by having shorter decision loops. The advantage will look quiet: fewer breakdowns, fewer bottlenecks, faster throughput, better use of assets, less chaos. But quiet advantages compound.

The pages ahead are written for the leader who refuses to approve another vanity pilot, and who also refuses to sit out the most important operating shift of this decade. The challenge is straightforward: pick one decision, build the smallest twin that can improve it, run it in shadow mode, prove the delta, and then scale with discipline.

Introduction: The Silent Revolution

"The next revolution will not look like magic. It will look like fewer surprises."

In recent months, something has shifted in boardrooms across industries. Executives who once competed to announce the most aggressive AI initiatives are now asking harder questions. The tone has moved from exuberance to scrutiny. Investment continues to grow, but so does the demand for proof. Although investment in AI continues to grow each year, the results of AI investments are not meeting expectations. The demonstrations were incredible. Pilot programs demonstrated great opportunity. Dashboards have never looked better. However, when the CFO asks for the P&L impact, the room falls silent.

The paradox at the center of today's enterprise AI is unprecedented enthusiasm and underwhelming outcomes. I refer to this as AI Fatigue: the growing gap between AI investment and demonstrable P&L impact, visible in delayed scaling decisions, pilot-to-production stalls, and increasingly pointed questions from CFOs. The symptoms appear to be spreading, particularly in industrial and regulated sectors where the path from insight to action is longest. CEOs that have championed transformational agendas are today downplaying the language, requesting "quick wins" rather than "moon shots," and quietly questioning if they were sold a vision of the future that no amount of technology can provide.

Ironically, they are correct to be skeptical, but for all the wrong reasons. The issue is not that AI does not function. Rather, the issue is that most organizations are applying AI to address the wrong problems. They built powerful telescopes to see what lies ahead but forgot to build the vehicles to travel there. Predictive models proliferated; the workflows, decision rights, and feedback loops required to act on predictions did not.

The Real Issue Is Not the Model

Most times, when organizations face issues with AI, they point the finger at the technology. Possibly the data was not clean enough. Maybe the algorithms require fine tuning. Or maybe the infrastructure cannot process the volume of data. These are easy to diagnose because they are technical, resolvable, and they do not impugn anybody's judgment. In over 150 transformation engagements I tracked, technical issues accounted for less than 20% of project failures. The rest traced to operating model misalignment.

The root cause of the issue is the operating model. Organizations are optimizing data while they should be optimizing decisions. Organizations are creating systems to inform them of what occurred last quarter rather than systems to advise them of what to do next Tuesday. Millions have been spent on infrastructure that creates beautiful charts depicting the operations described in the charts and simultaneously hemorrhaging money due to inefficiencies, downtime, and misallocated resources. Most organizations now capture more operational signals than they can convert into timely action. The constraint is rarely data availability. It's the architecture that connects data to decisions within the required time window.

I refer to this phenomenon as "Analytics Theater": The organizational performance of being perceived as data-driven while actually making decisions in the same way as it ever was. In Analytics Theater, dashboards multiply, data lakes grow, and executive presentations contain more sophisticated visualizations. However, the frontline manager still relies on intuition. The maintenance technician still performs preventive maintenance in the same manner as before, regardless of sensor data, because even when they notice an anomaly, they lack both the diagnostic context to prove their instinct correct and the authority to act without approval. The supply chain planner still develops the same buffers in his planning since nobody believes the forecast. Everybody has access to more information than ever; however, the information rarely influences what anyone does. The problem is not just visibility. It is that visibility without authority changes nothing.

There's a persistent gap between the knowledge that organizations possess and the actions they take. Throwing additional AI at the knowledge side of the equation will not bridge this chasm. Analysis alone does not create operational excellence; improvement requires decision rights, workflows, and action. At some point, some type of action needs to occur in the physical world and that is typically where most AI programs get stuck: the transition from insight to action.

A Different Type of Mirror

The Mirror Strategy begins with a simple premise: advantage comes from decision loops, not reports. A Digital Twin, as defined here, is not a prettier dashboard and not a cinematic 3D model built for a keynote. It is a living operational model, kept in sync with reality, that converts that synchronization into executable decisions.

A useful analogy is navigation. A paper map informs. A GPS guides. The map can be beautifully designed and still be wrong at the moment it matters. A twin, at its best, behaves like GPS: it tracks the current state, anticipates what is likely to happen next, and recommends a route that respects constraints.

The First Twin Was Manual

The idea is older than the phrase. Long before vendors marketed "Digital Twins," NASA was already practicing the underlying discipline: keep a working mirror of a complex system, keep it current, and use it to decide what to do next.

In 1970, computers were not powerful enough to simulate the full physics of a spacecraft in real-time software. So, NASA leaned on what engineers now call Hardware-in-the-Loop: physical cockpits on the ground, wired into computers, behaving as stand-ins for the real vehicle in space. In Houston, NASA maintained multiple simulators—not just screens, but tangible panels, switches, and gauges connected to mainframes.

The Twin: A ground-based cockpit-and-computer system that could be configured to match the spacecraft's condition.

The Data Link: The link was not automated telemetry streaming into a model. It was manual transmission: astronauts reading instrument values over the radio while engineers dialed those values into ground-based simulators to mirror the crippled ship's state.

The Feedback Loop: Engineers tested procedures and the startup sequence on the ground first, and only then read the steps back to the crew.

Verdict: It was a manual twin. The "API" was a human being talking on a radio.

The Timeline of the Name

The practice started decades earlier. The marketing arrived later.

1970 (The act): NASA uses mirrored ground systems to rehearse solutions and help bring Apollo 13 home.

2002 (The concept): Dr. Michael Grieves (University of Michigan) formalizes the "Mirrored Spaces" model in the context of Product Lifecycle Management.

2010 (The name): John Vickers (NASA) and Grieves popularize the term "Digital Twin" through NASA road mapping work.

Between Apollo and Grieves' formalization, industrial simulation matured through CAD/CAM systems in the 1980s and SCADA-based process control in the 1990s. These technologies laid the groundwork, but lacked the continuous synchronization and decision-loop architecture that define the modern twin.

Modern twins usually keep themselves current through the Internet of Things: The living nervous system of the physical world. Cameras, vibration sensors, pressure gauges, flow meters, and countless other devices stream the signals that let a model stay honest. This book will not dive into the sensor technology itself. It is not a book about hardware. It is a book about strategy: what to twin, why it matters, and how to turn synchronization into advantage.

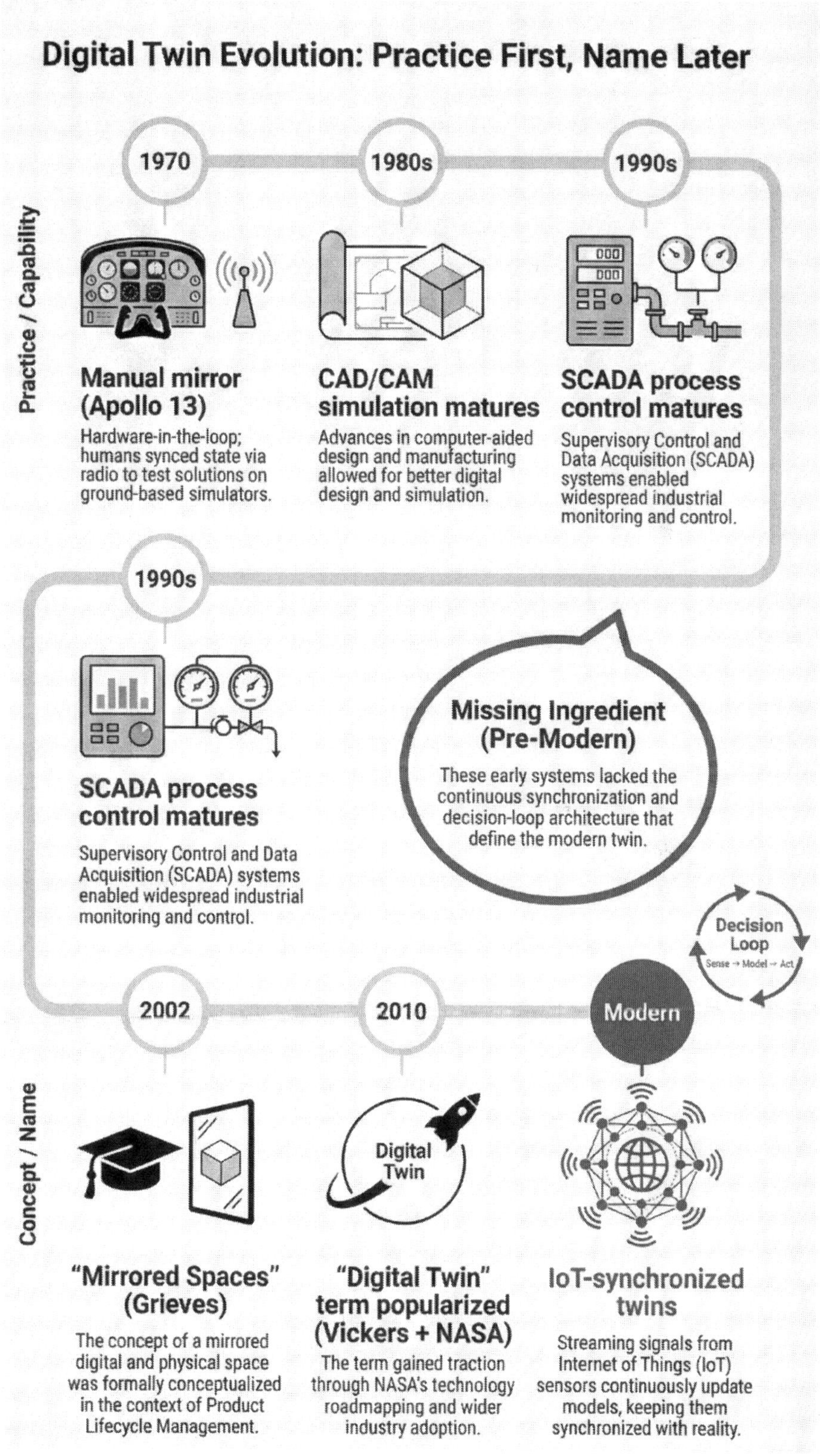
Digital Twin Evolution: Practice First, Name Later
Practice / Capability
1970
1980s
1990s
Manual mirror (Apollo 13)
Hardware-in-the-loop; humans synced state via radio to test solutions on ground-based simulators.
CAD/CAM simulation matures
Advances in computer-aided design and manufacturing allowed for better digital design and simulation.
SCADA process control matures
Supervisory Control and Data Acquisition (SCADA) systems enabled widespread industrial monitoring and control.
1990s
SCADA process control matures
Supervisory Control and Data Acquisition (SCADA) systems enabled widespread industrial monitoring and control.
Missing Ingredient (Pre-Modern)
These early systems lacked the continuous synchronization and decision-loop architecture that define the modern twin.
Decision Loop
Sense → Model → Act
Concept / Name
2002
2010
Modern
Digital Twin
"Mirrored Spaces" (Grieves)
The concept of a mirrored digital and physical space was formally conceptualized in the context of Product Lifecycle Management.
"Digital Twin" term popularized (Vickers + NASA)
The term gained traction through NASA's technology roadmapping and wider industry adoption.
IoT-synchronized twins
Streaming signals from Internet of Things (IoT) sensors continuously update models, keeping them synchronized with reality.

From Theater to Operations

The promise of this book is relatively straightforward, although quite ambitious: I will demonstrate how to progress from Analytics Theater to operational environments that generate first measurable impact in 30 to 60 days on a single decision, with larger ROI compounding over quarters as the approach scales. This is not a theoretical exercise. The organizations I will present in this book include manufacturers like General Motors that decreased unplanned downtime by 25% in stamping operations, hospitals like Queen's Health Systems in Hawaii that reduced average length of stay, unlocked the equivalent of an entire wing of capacity without laying a single brick, and fundamentally changed how decisions flow through the organization by empowering charge nurses and bed managers to act within validated guardrails rather than waiting for administrative approval, utilities like Thames Water that documented an 18 million liter daily reduction in water loss by detecting leaks invisible to traditional inspection methods. These organizations have not obtained these results through some revolutionary technology beyond your financial capability. They have achieved them by considering a different perspective on what AI should accomplish.

The secret is not in the algorithm. It is in the architecture. It is in designing systems wherein the output is not a report to be reviewed but a recommendation to be executed or, increasingly, an action that executes within carefully defined guardrails. It is in developing what I refer to as decision loops: closed circuits where data flows in, the model analyzes it, a prescription emerges, the operation responds, and the results feed back into the model to enhance it.

Throughout the subsequent chapters, I will detail how to construct these decision loops. First, we will assess why most AI programs fail to advance past the dashboard level of operational maturity and what is necessary to cross that threshold. Next, I will provide a structured methodology commonly referred to as the AI Maturity Ladder that identifies where your organization currently resides and what capabilities you must obtain to ascend. Subsequently, we will explore the architecture of a functional Digital Twin layer by layer, so you comprehend the requirements without getting bogged down in technical terminology. Finally, I will outline a tested playbook for initiating a

Minimum Viable Twin in thirty to sixty days: A focused proof of concept that generates tangible value while you learn.

Most importantly, we will confront the difficult inquiries that executives pose when autonomy is introduced into the discussion. How do you preclude a probabilistic model from generating deterministic errors? What guardrails assure the system understands when to suggest and when to escalate? How do you maintain accountability when decisions are made quicker than a person can evaluate them? These are not obstacles to be dismissed. These represent the regulatory hurdles that distinguish toys from tools and pilots from production.

The Quiet Revolution

In short, while the technology press focuses on chatbots and generative AI, technologies that are both incredibly impressive and largely unrelated to operational improvement, a quiet revolution is transforming the manner in which industrial organizations operate. It is occurring in factories where predictive models initiate maintenance before a failure occurs. It is occurring in hospitals where patient flow is managed by algorithms that can observe constraints that humans do not perceive. It is occurring in energy grids, water networks, and logistics hubs where the distance between the physical environment and the digital representation thereof is diminishing to nearly zero.

The organizations driving this revolution are not generally the largest IT spenders or the most advanced data science teams. They are the organizations that have learned to inquire regarding a different query. Not "What can AI do to improve our reports?" but "What decisions generate the greatest cost or risk, and how can we close the decision loop from insight to action?" This represents a slight shift in perspective, but it radically alters every aspect of how you design, construct, and regulate your systems.

There is a corollary question that proves equally transformative: Who should be making this decision? In most organizations, small problems escalate into large crises because the people who first notice them lack the information or authority to resolve them at the source. A vibrating motor becomes a seized motor becomes a missed shipment becomes an executive crisis call. A well-designed twin interrupts this escalation by giving frontline workers both the diagnostic context and the validated recommendations they need to act with confidence. The result is an organization that functions more like a healthy immune system, neutralizing threats at the cellular level before they can spread. Executives stop firefighting and start architecting. That shift alone can justify the investment.

If that question resonates with you and if you have experienced the frustration of expending resources on analytic capabilities that appear to produce little to no ROI, then this book should be your roadmap. Prior to proceeding, we must determine why so many organizations continue to be mired in the initial stages. We must explore the pitfall that snares even the most sophisticated organizations: the tantalizing, costly and ultimately hollow promises of the dashboard. That is where we will begin.

Chapter 1 – The Dashboard Trap

"Visibility is not control. A dashboard can inform the driver, but it cannot steer."

The plant supervisor, a veteran of thirty years named Bill, felt a familiar rhythm in the quiet hum of the factory floor. He was watching a wall of green lights on his indicators for uptime, yield, and order fill rate and all appeared to be operating as expected. Suddenly, one of the pumps failed, and within minutes the production line came to a halt, the entire shipping schedule was disrupted, and the dashboard turned completely red. The display was correct. It was just telling the truth a bit late.

Dashboards have an underlying power to make complicated systems understandable. Dashboards are gauges for managers. They're simple, shared measures that direct attention toward the appropriate focus area. Dashboards help teams understand the results of their previous day's work and determine if a prior week's correction action was successful in moving the indicator in the correct direction. They allow "what-if" scenarios to be discussed based on agreement on today's numbers.

A gauge can measure altitude, but it has no ability to predict the next pocket of turbulence nor provide a better route through turbulent air. Now consider a dashboard with added leading indicators (predictive KPIs) estimating what will occur next with recommended corrective actions for each issue. At this time, the dashboard changes from being a rear-view mirror to being a front window. In doing so, the inevitable question arises: what could that do to a business?

Dashboards provide the foundation: baseline metrics, accountability, and the feedback loop that confirms whether actions worked. A twin builds on this foundation but crosses a critical threshold. A twin begins when three conditions are met: (1) an explicit model of constraints and system behavior exists, (2) outputs route into a decision engine rather than merely a display, and (3) prescriptions flow into operational workflows with clear ownership and feedback. Without all three, you have advanced analytics. With all three, you have a twin.

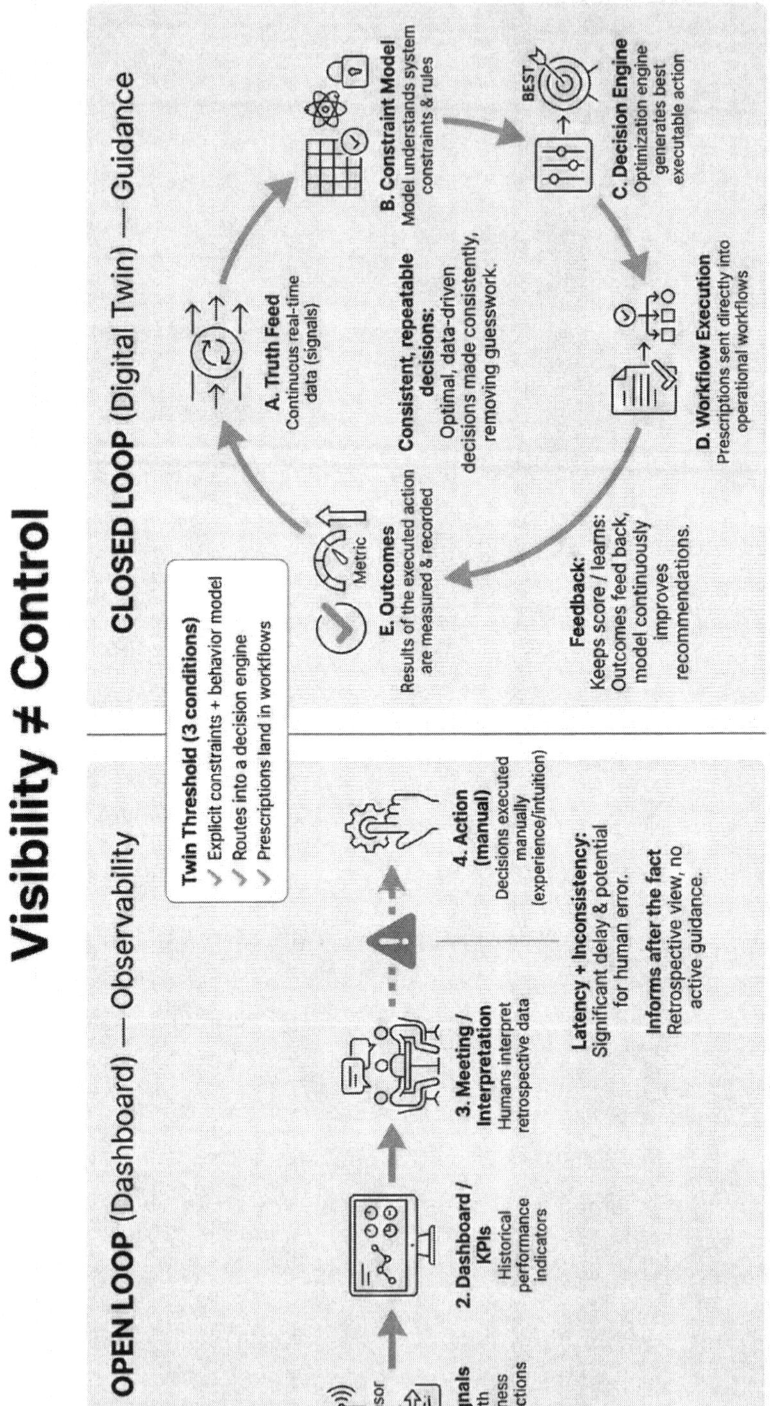
Visibility ≠ Control
OPEN LOOP (Dashboard) — Observability
CLOSED LOOP (Digital Twin) — Guidance
Sensor
1. Signals with Business transactions
2. Dashboard / KPIs Historical performance indicators
3. Meeting / Interpretation Humans interpret retrospective data
4. Action (manual) Decisions executed manually (experience/intuition)
Latency + Inconsistency: Significant delay & potential for human error.
Informs after the fact Retrospective view, no active guidance.
Twin Threshold (3 conditions)
Explicit constraints + behavior model
Routes into a decision engine
Prescriptions land in workflows
A. Truth Feed Continuous real-time data (signals)
B. Constraint Model Model understands system constraints & rules
BEST
C. Decision Engine Optimization engine generates best executable action
D. Workflow Execution Prescriptions sent directly into operational workflows
Metric
E. Outcomes Results of the executed action are measured & recorded
Consistent, repeatable decisions: Optimal, data-driven decisions made consistently, removing guesswork.
Feedback: Keeps score / learns: Outcomes feed back, model continuously improves recommendations.

Two Layers, Two Worlds

To appreciate how both dashboards and Digital Twins can be of use to organizations, it will help to distinguish between the two layers that exist within all organizations (the reporting layer and the operating layer).

The reporting layer is the layer of business intelligence, analytics, and executive review. It is retrospective in nature and responds to questions like "what happened" and "how did we do." The reporting layer is the place where decisions about governance, learning and accountability are made. This is also where dashboards and reports are most successful at creating visibility into performance, comparisons and explanations of that performance.

The operating layer is the layer of production, logistics and real time operations. The operating layer is prospective in nature and responds to questions like "what should we do next?" and "what should we change right away?" The operating layer is where the product is manufactured, patient is treated, shipment is made and maintenance is done. The operating layer generates revenue and consumes expense.

Today, in most organizations, the biggest investments have been in the reporting layer: Better data, clearer visuals, faster refresh rates. In contrast, the operating layer continues to rely on meetings, phone calls, and experienced judgment to convert signals into action. This is not a mistake; it is simply the natural stopping point for dashboards.

TWO LAYERS, TWO WORLDS

Dashboards explain the past. Twins shape the next decision.

LAYER 1: THE REPORTING LAYER (WORLD OF DASHBOARDS)

Explains The Past

Focuses on retrospective analysis of historical performance.

Answers: 'What happened?'

Provides visibility into past performance for learning and governance.

A Rear-View Mirror

Illuminates the system but cannot steer future outcomes.

Business Intelligence & Reports

Creates visibility through dashboards, charts, and executive reviews.

LAYER 2: THE OPERATING LAYER (WORLD OF DIGITAL TWINS)

Shapes The Future

Focuses on prospective decisions and real-time execution.

Answers: "What should we do next?"

Generates revenue and consumes expenses through direct action.

A Guidance System (GPS)

Provides prediction and direction to navigate what's next.

Digital Twins & Decision Engines

Links predictive indicators to executable, constraint-aware actions.

Dashboards illuminate the system. However, to improve the future of the organization, the organization must include what dashboards are not capable of providing independently: Prediction and direction.

When leaders feel that they have failed with AI, the underlying reason for this feeling is typically the difference between viewing and guiding. Therefore, the opportunity is not to abandon dashboards, but to expand upon them and add predictive KPI's that are able to see into the future, and link those indicators to corrective actions that take into account limitations. As such, the dashboard develops from an instrument panel into a guidance system.

The Latency Tax

Another problem with the dashboard-oriented approach is the latency between the physical world and the organizational awareness of the physical world. In the vast majority of organizations, the time it takes for data to flow from sensors to systems to warehouses to dashboards to meetings to decisions is sufficient that the world has moved on before the organization realizes what occurred. The machine that displayed symptoms of failure last Tuesday has likely failed by today. The shipment that was identified as being at risk has likely arrived late. The patient who required timely intervention has likely been discharged.

I refer to this phenomenon as the "Latency Tax": the indirect costs organizations incur due to their delayed understanding of the physical world. The Latency Tax manifests in reactive firefighting rather than pro-active prevention, in missed opportunities that were clearly identifiable in the data, but opaque in terms of the workflow, in decisions made based on stale information dressed in real-time garb. A dashboard that updates every 15 minutes still presents the world as it was 15 minutes ago, and in rapidly changing environments, that delay can be the difference between identifying a problem and addressing it, versus cleaning up after the problem has occurred. Multiply that across hundreds/thousands of decisions each day, and the cumulative costs become enormous.

The Latency Tax creates an illusion of control. A screen that refreshes every fifteen minutes feels "real-time," yet it still presents the world as it was fifteen minutes ago. Recency is not the same as relevance. Knowing the temperature was X degrees 15 minutes ago does not help prevent the temperature from exceeding its threshold 8 minutes from now. By the time the dashboard updates, the excursion has already occurred, or worse, already cascaded into a quality defect or equipment trip. The data was available to record at the appropriate time, but it was not available to usefully inform an immediate decision.

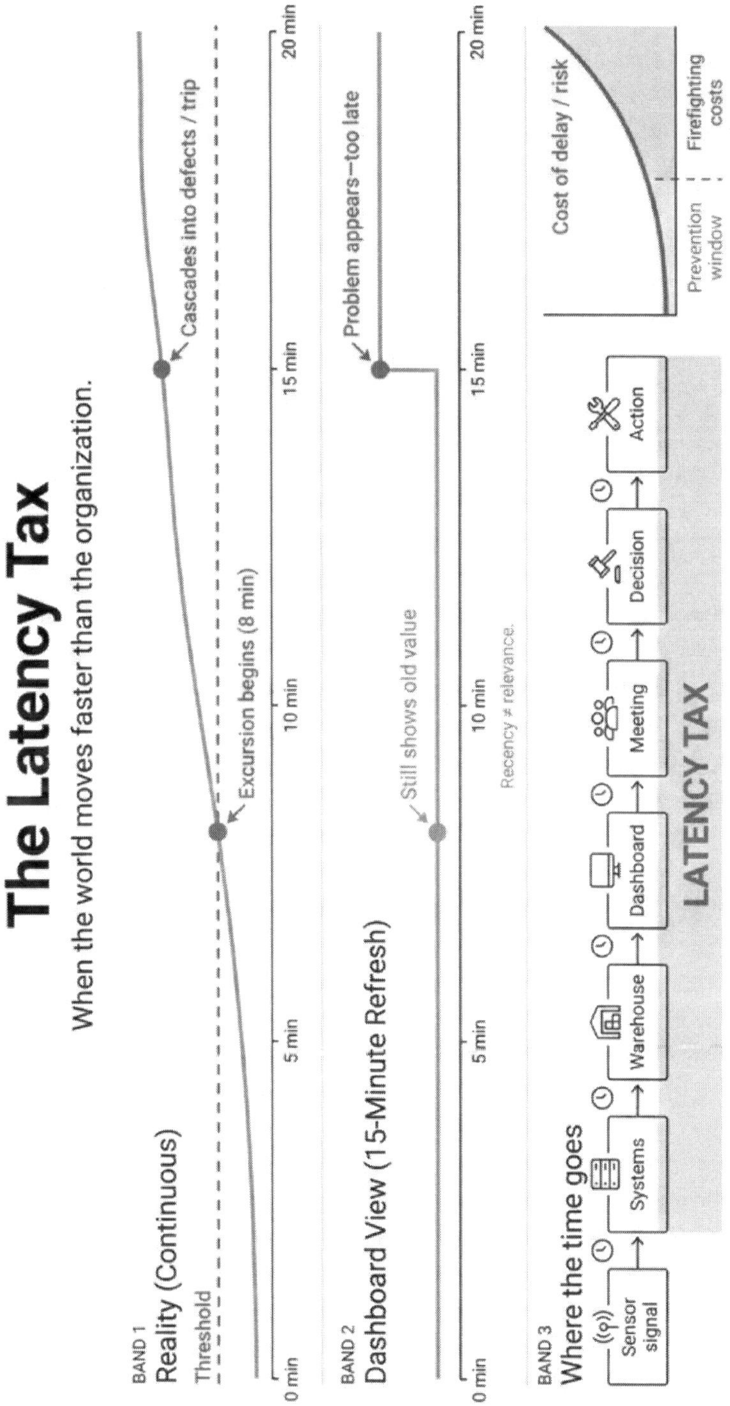
The Latency Tax
When the world moves faster than the organization.
BAND 1
Reality (Continuous)
Threshold
Excursion begins (8 min)
Cascades into defects / trip
0 min
5 min
10 min
15 min
20 min
BAND 2
Dashboard View (15-Minute Refresh)
Still shows old value
Problem appears—too late
0 min
5 min
10 min
15 min
20 min
Recency ≠ relevance.
BAND 3
Where the time goes
Sensor signal
Systems
Warehouse
Dashboard
Meeting
Decision
Action
LATENCY TAX
Cost of delay / risk
Prevention window
Firefighting costs

Where Does This Leave Us?

Companies have invested considerable sums in establishing data infrastructure. i.e., the pipes that carry information from the physical world into digital systems. Companies have invested in analytics platforms – the software tools that convert raw data into graphs and metrics. But most companies have not invested in the bridge that connects the above capabilities to operational workflows. The bridge that converts an insight into an action. The bridge that closes the loop between knowing and doing.

That is precisely what a Digital Twin serves as. Not as a fancy visualization or a 3-D model, but as a decision-making tool that sits between the data infrastructure and the operational workflow. The twin receives inputs from the physical operation, uses models that reflect the constraints and goals of the physical operation to interpret the inputs, generates prescriptions for what should occur next, and supplies the prescriptions to the workflow(s) where action occurs. The twin does not simply report on the state of the physical world, it assists in defining the future of the physical world.

However, before we can develop that bridge, we need to agree on what we are attempting to build. The term "Digital Twin" has been used, abused, and distorted so widely by both vendors and consultants that it risks becoming meaningless. When I use the term "Digital Twin", I am referring to something very specific, tangible, and quantifiable. In the next chapter, we will describe what I am referring to in terms of a Digital Twin, and equally important, what I am not referring to.

Chapter 2 – Defining the Mirror

"A twin becomes useful when it argues with reality every day and keeps score."

A Digital Twin is defined as a living operational model, that is continually synchronized with physical realities and then translates that synchronization into actionable decisions. Each term is intentional:

- **Living**: A Digital Twin is called "living" since it continues to evolve as physical realities continue to change. It's a dynamic model, unlike a static model that's built once and referenced ad infinitum.
- **Operational**: A Digital Twin is referred to as "operational," and is focused on process and decision-making, not aestheticism nor documentation.
- **Synchronized**: A Digital Twin is described as "synchronized," as it maintains a real-time, or nearly real-time, connection to physical reality via sensors, transactional activity, etc.
- **Executable**: A Digital Twin provides "executable" decisions (not reports, insights, recommendations stuck in inboxes) and provides those decisions directly into operational workflows.

The Four Pillars of a Digital Twin

A living operational model, synchronized to reality, that produces executable decisions.

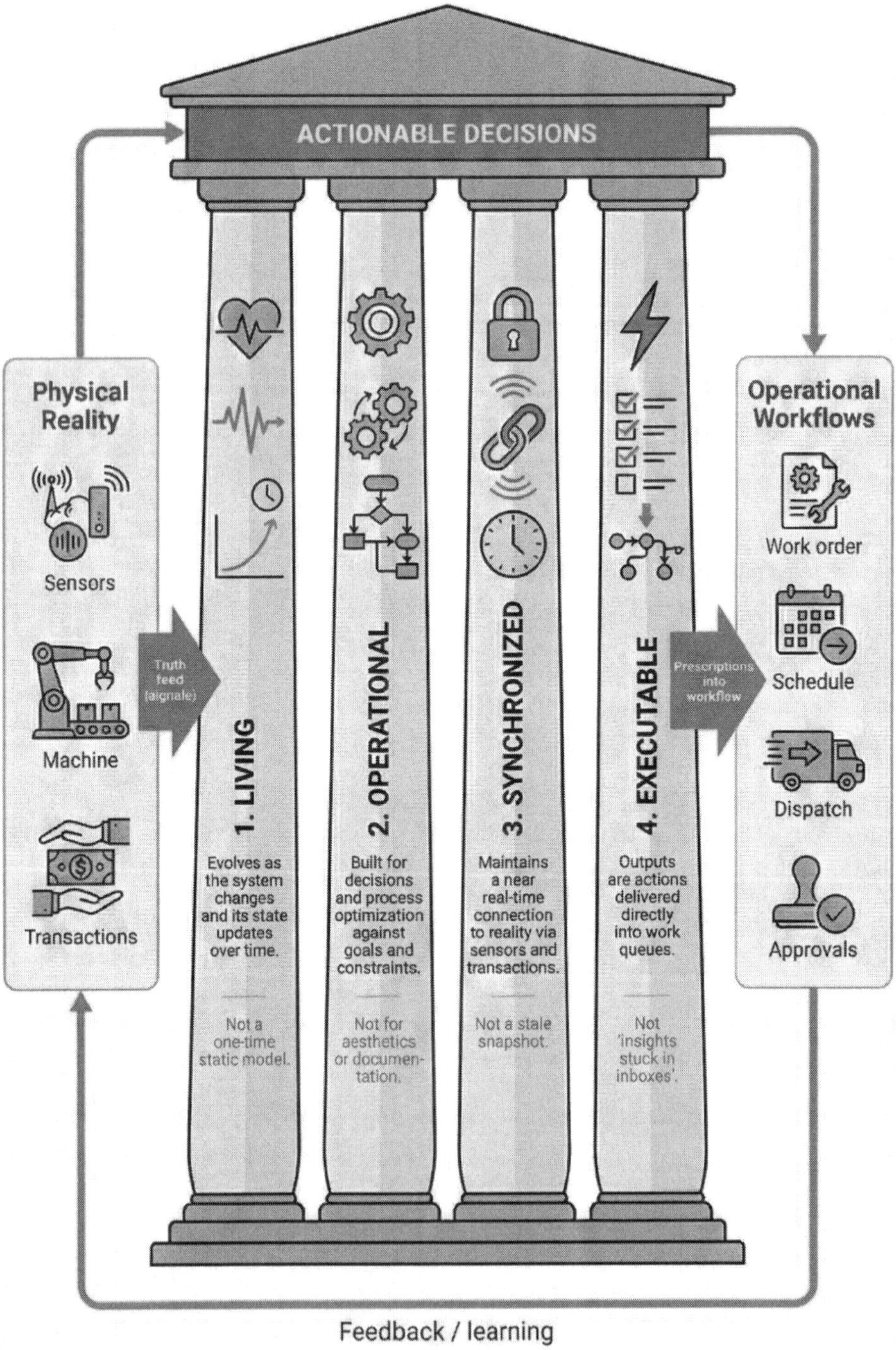

Navigation Analogy

As introduced earlier, the map-to-GPS analogy captures the core distinction. The map is essential: You cannot navigate without knowing the roads. But the GPS adds real-time position, traffic awareness, predictive routing, and course correction when conditions change. A dashboard provides the map. A twin provides the GPS: synchronized state, predicted outcomes, and recommended action.

What a Digital Twin Isn't

There is considerable confusion regarding Digital Twins, largely due to the marketing efforts of various vendors attempting to sell their products as Digital Twins. To help clarify what a Digital Twin is and what a Digital Twin is not, the following is a list of examples of what a Digital Twin is not:

Firstly, a Digital Twin is not a 3D model. When most people hear the term Digital Twin, they envision a 3D model of their production floor, or a turbine spinning in a virtual reality environment. While 3D models may be visually appealing, and even realistic, geometry does not equate to decision making advantage. For example, building a highly detailed 3D model of a company's supply chain will never inform the decision of whether to ship product through Memphis or Dallas when inclement weather occurs. The visuals are window dressing; the value resides in the underlying logic.

Secondly, a Digital Twin is not a chatbot (though the two work well together). A chatbot is an interface; a twin is the underlying model, decision engine, and workflow integration. A chatbot can tell you what a company's maintenance schedule states. A twin can analyze vibration signatures, correlate them with past failures, and recommend the optimal service window. A twin may be queried through a conversational interface, but the interface is not the twin. The value lies in the calculation and the closed loop, not in the conversation.

Thirdly, a Digital Twin is not a one-time simulation (though it may use simulation as a tool). Traditional simulations answer a specific question: What happens if we add a second shift? What is the maximum throughput this line can sustain? Once answered, the analysis closes. A twin differs in two ways: continuous synchronization with the physical system, and an operational decision loop that acts on what the model learns. The twin may run thousands of simulations daily, but the distinguishing feature is not simulation frequency; it is the closed loop that converts model outputs into workflow actions and feeds results back to improve the model.

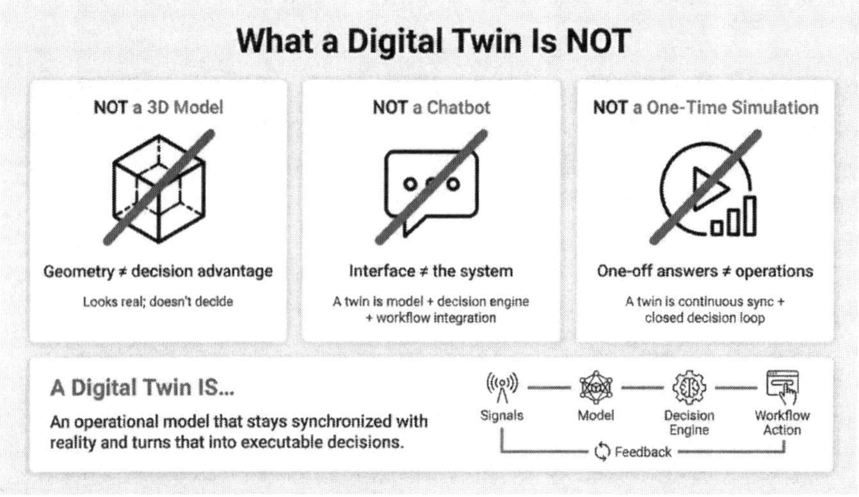

Temporal Connectivity

The ability of a Digital Twin to maintain a continuous and evolving relationship between its digital model and its physical counterpart, over time, is referred to as temporal connectivity. Temporal connectivity is not a static image of a piece of equipment at a point-in-time. Temporal connectivity is a continuous flow of information from the physical world into the digital model, allowing the digital model to evolve as the physical world changes.

For example, a pump in a manufacturing plant is initially modeled as a new pump operating at peak efficiency. Six months later, after many hours of operation, the pump has aged and worn. The vibration signature of the pump has changed slightly, as well as the energy consumed by the pump. If a static model were used to model the pump, it would likely be assumed to operate at peak efficiency, even though the pump has significantly degraded. However, if a Digital Twin model were used, it would have monitored the pump's degradation, incorporated that degradation into the model, and updated the recommendations accordingly. The Digital Twin would understand that this is no longer the same pump it was six months ago, despite the pump looking the same from the outside.

Temporal connectivity allows a Digital Twin model to transform itself from a simple reference document to a decision-making partner. The Digital Twin model does not simply represent the physical asset; it understands the physical asset in the context of its operation. The Digital Twin model can predict how the physical asset will react to various conditions because it has been observing the physical asset over time, learning from the physical asset, and continuing to update its knowledge base.

Now that you have a clear understanding of what a Digital Twin is, and what a Digital Twin is not, we can begin to evaluate where your organization currently stands in terms of developing a Digital Twin. Most organizations are somewhere between the beginning of the journey and the destination. Few organizations will be ready to deploy a completely autonomous Digital Twin. Instead, organizations will follow a progression, or a maturity curve, that will need to be climbed. In the next chapter, we will outline this maturity curve and provide a tool for evaluating your organization's readiness to develop and deploy a Digital Twin.

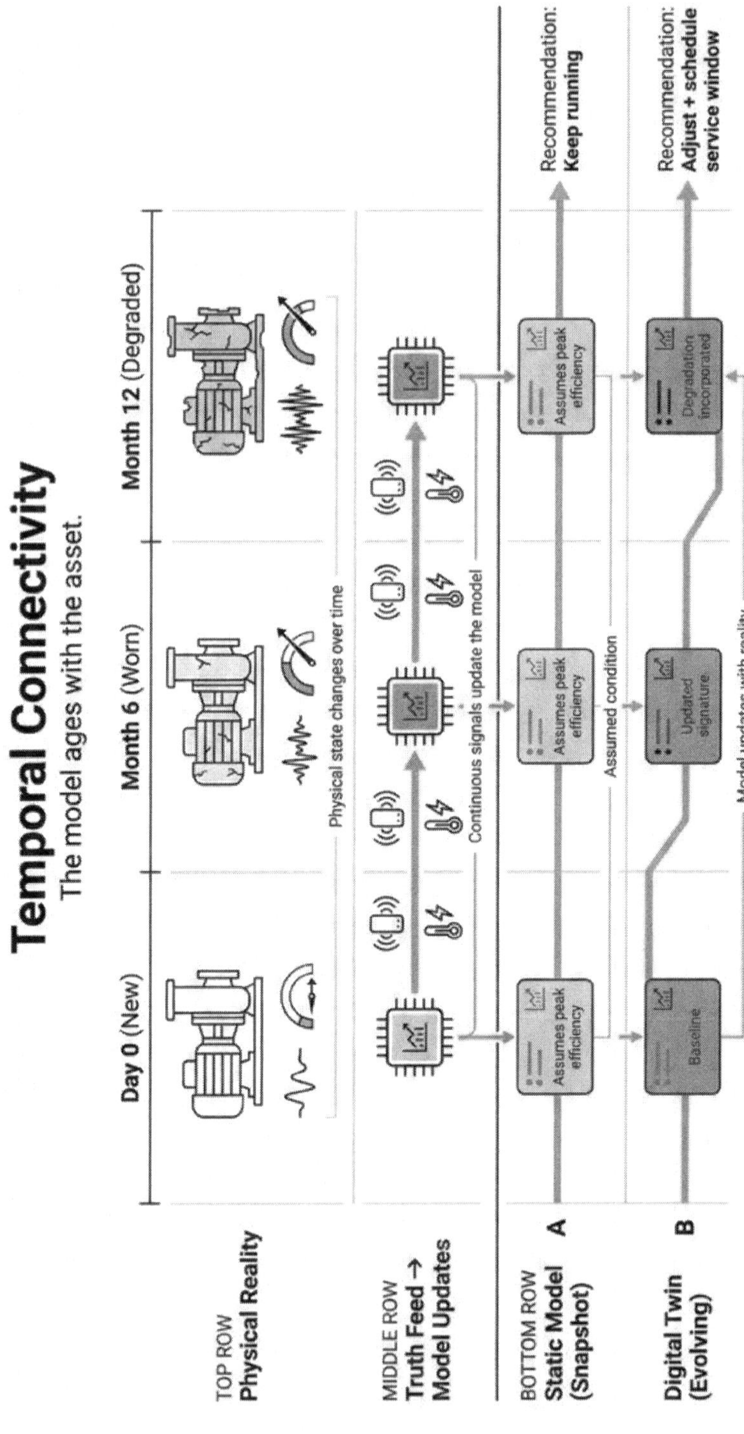
Temporal Connectivity
The model ages with the asset.
TOP ROW
Physical Reality
MIDDLE ROW
Truth Feed → Model Updates
BOTTOM ROW
Static Model (Snapshot)
A
Digital Twin (Evolving)
B
Day 0 (New)
Month 6 (Worn)
Month 12 (Degraded)
Physical state changes over time
Continuous signals update the model
Assumes peak efficiency
Assumes peak efficiency
Assumes peak efficiency
Assumed condition
Baseline
Updated signature
Degradation incorporated
Model updates with reality
Recommendation: Keep running
Recommendation: Adjust + schedule service window

Chapter 3 – The AI Maturity Ladder

"Maturity is not smarter answers. It is faster, safer decisions."

Many executives generally underestimate how close or far they are to achieving Operational AI Maturity and often overestimate the sophistication of their current capabilities. Most confuse having data with having insight and having insight with taking action. To clarify the above, let's introduce the AI Maturity Ladder which consists of five distinct levels of progression, ranging from passive reporting (lowest) to semi-autonomous operations (highest).

At each level there is a unique question that is answered and each question is more valuable than the last. Being able to clearly identify where your organization resides on the Ladder is the first step in climbing it.

Level 1: The Reflective – What happened

The first level provides the answer to the least valuable of the five questions. At this level, the answer to the question is: "**What Happened?**" This is the world of Descriptive Analytics: Dashboards, reports, visualizations, etc., which provide a view of historical performance. At this level, you can see that machine uptime was 87% last month. You can see that customer complaints rose 15% in Q3. You can see that warehouse throughput declined during the holiday rush.

Many organizations remain effectively stuck at this level, including some that believe they have progressed beyond it. They have invested in business intelligence platforms and hired analysts, yet the work still ends in visibility. The typical constraint is not reporting capability, but execution capability: unclear decision ownership, workflows that cannot absorb recommendations, and a lack of feedback loops that prove whether actions worked.

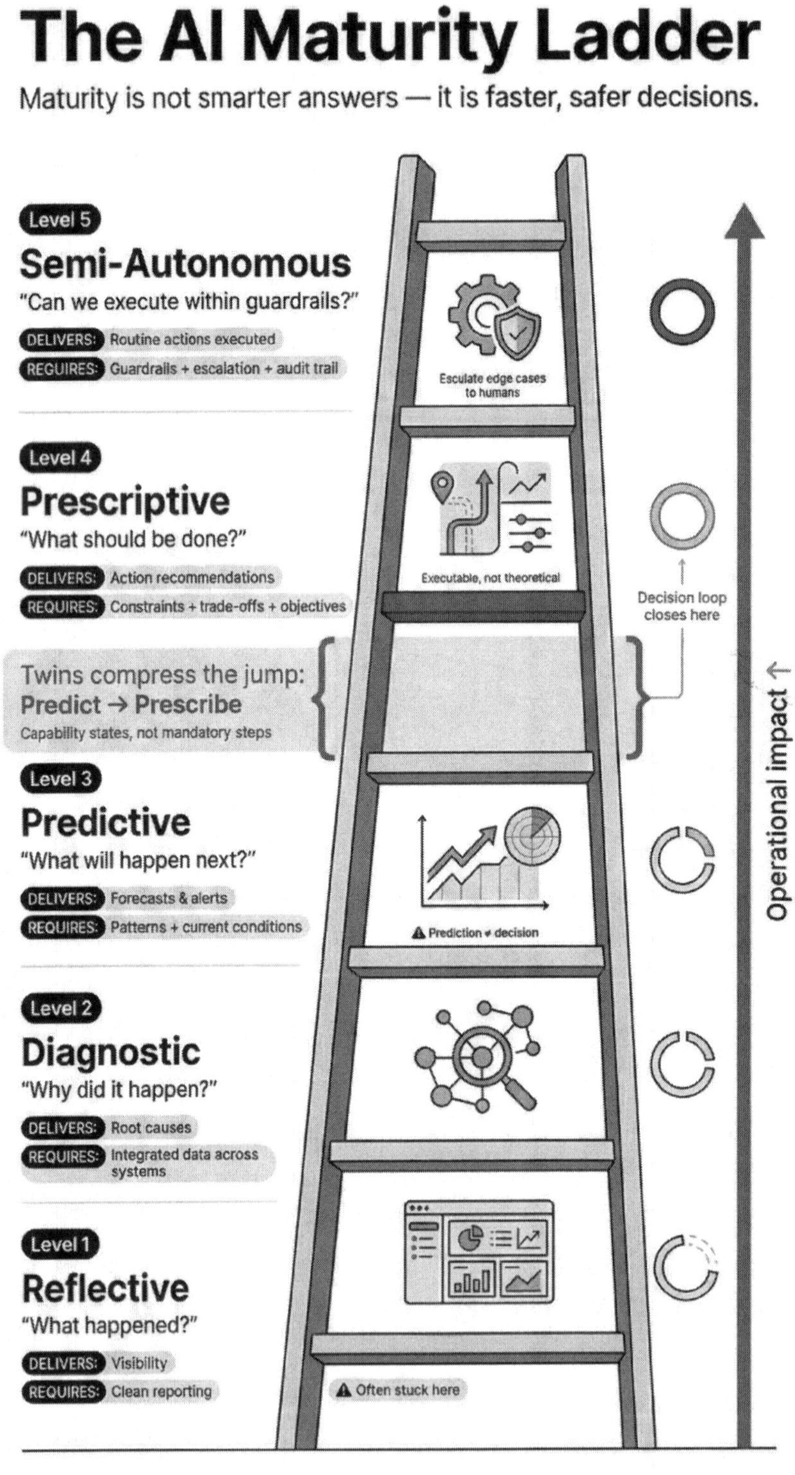
The AI Maturity Ladder
Maturity is not smarter answers — it is faster, safer decisions.
Level 5
Semi-Autonomous
"Can we execute within guardrails?"
DELIVERS: Routine actions executed
REQUIRES: Guardrails + escalation + audit trail
Esculate edge cases to humans
Level 4
Prescriptive
"What should be done?"
DELIVERS: Action recommendations
REQUIRES: Constraints + trade-offs + objectives
Executable, not theoretical
Decision loop closes here
Twins compress the jump:
Predict → Prescribe
Capability states, not mandatory steps
Level 3
Predictive
"What will happen next?"
DELIVERS: Forecasts & alerts
REQUIRES: Patterns + current conditions
Prediction ≠ decision
Level 2
Diagnostic
"Why did it happen?"
DELIVERS: Root causes
REQUIRES: Integrated data across systems
Level 1
Reflective
"What happened?"
DELIVERS: Visibility
REQUIRES: Clean reporting
Often stuck here
Operational impact →

Level 2: The Diagnostic – Why did it happen?

The Second Level asks a question that is more valuable than the first. The question asked is: **"Why Did It Happen?"** This is Diagnostic Analytics – the ability to drill down to the root cause and identify the drivers of outcomes. Not only did machine uptime decline, but it declined because of bearing failures in three specific units that were all installed during the same maintenance window. Not only did complaints increase, but they increased because shipping delays from a single distribution center cascaded through the network.

Achieving Level 2 capability requires more than better dashboards. It requires integrated data that connects effects to causes across systems. Most organizations that claim to be here are actually toggling between Level 1 dashboards and manual investigation, with humans doing the diagnostic work that the system cannot do. True Level 2 capability means the system surfaces the "why" without requiring an analyst to hunt for it.

Level 3: The Predictive – What Will Happen Next?

The Third Level changes perspective from retrospective to prospective. The question asked is: "**What Will Happen Next?**" This is Forecasting. i.e., using historical patterns and current conditions to project future states. The system does not just report that the pump is showing elevated vibration; it predicts that, based on the trajectory, the pump will likely fail within the next seventy-two hours. It does not just note that inventory is declining; it forecasts a stockout in two weeks if consumption continues at its current rate.

Level 3 is where machine learning typically enters the picture. Prediction at scale requires pattern recognition beyond human capacity. However, prediction alone, while valuable, still leaves a critical gap. Knowing that something will happen does not tell the organization what to do about it. The maintenance manager who learns a pump will fail in 72 hours still has to decide: Repair now, schedule for tomorrow, run to failure and have a backup ready? The forecast creates awareness but not action.

Level 4: The Prescriptive – What Should Be Done?

The Fourth Level closes the above identified gap. The question asked is: **"What Should Be Done?"** This is Prescriptive Analytics: Optimization that considers constraints, trade-offs and objectives to recommend specific actions. The system does not just predict pump failure; it analyzes the production schedule, spare parts availability, technician capacity and downstream impact to prescribe: "*Schedule maintenance for the Tuesday night shift when line 3 is already down for changeover, minimizing total production loss.*"

Prescription is the sweet spot of the Digital Twin. Level 4 requires exactly the kind of living operational model defined in the previous chapter. I.e., one that understands not just what's happening but how the entire system interconnects. The model must know the constraints (we can't shut down line 1 during peak demand). The model must know the objectives (minimize total cost of maintenance while maintaining throughput). And the model must know the current state of all relevant variables. Only then can it generate a prescription that is actually executable rather than theoretically optimal.

Level 5: The Semi-Autonomous – Can We Execute Within Guardrails?

The Fifth and final Level of the ladder answers a different type of question: "**Can We Execute Within Guardrails?"** This is Controlled Autonomy: Systems that don't just recommend actions but take them, within carefully defined boundaries. The Twin doesn't just prescribe the maintenance schedule; it updates the work order system, notifies the technicians, reserves the parts, all without human intervention, because the decision falls within parameters that humans have pre-approved.

Level 5 isn't about removing humans from the loop; it is about redefining their role. The system handles routine decisions autonomously while escalating edge cases, anomalies, and high-stakes choices to human judgment. The human becomes the supervisor rather than the operator, intervening when needed rather than executing every step. This shift requires deliberate change management. Some operators will experience it as elevation; others may perceive it as demotion from skilled decision-maker to passive monitor. Organizations that succeed at Level 5 invest as heavily in role redesign and training as they do in technology.

How Twins Help the Climb

Here's the strategic insight that most organizations miss: You don't climb the ladder one rung at a time through incremental improvement. A well-designed twin builds Level 3 and Level 4 capabilities simultaneously. The ladder describes capability states, not required steps. The twin automates prediction (Level 3) to present users directly with prescription (Level 4). When data is accessible, the decision is narrow, and integrations are limited, teams can reach Level 4 for a single decision in 60 to 120 days. The rigor of building a twin, i.e., defining constraints, feedback loops, and decision logic upfront, collapses years of gradual capability building into months of focused development.

However, there is a subtler transformation embedded in this progression that the ladder alone does not capture. Notice what happens as organizations climb: decisions that once required senior judgment begin flowing to the people closest to the work. The maintenance technician who once escalated a vibration anomaly to a supervisor now receives a validated prescription directly. The charge nurse who once waited for administrative approval now acts within preapproved guardrails. The twin does not merely improve decisions; it redistributes where decisions are made. This redistribution carries profound implications for how organizations operate, how frontline workers experience their roles, and perhaps most importantly, how executives spend their time. Before we examine the technical architecture that makes twins function, we need to understand this organizational dividend, because it may be the most undervalued benefit of the entire endeavor.

Chapter 4 – The Delegation Dividend

"The best executive is the one who has sense enough to pick good people to do what needs to be done, and self-restraint enough to keep from meddling with them while they do it."□

There is a story I think about often when executives tell me they are hesitant to push decision making authority downward in their organizations. It involves a factory floor in Connecticut, a motor that was beginning to vibrate, and a worker who noticed it three days before anyone with authority to act finally did something about it.

The worker heard the vibration on a Monday. It was subtle, the kind of anomaly that experienced operators learn to notice in the hum of machinery. But the worker lacked two things: the data to diagnose the root cause and the authority to stop the line. So, the vibration continued. By Wednesday, the motor's performance had degraded noticeably. The worker mentioned it to a supervisor, who mentioned it to a maintenance manager, who scheduled a review for Friday. On Thursday afternoon, the motor seized. The line stopped. A key shipment was missed. A week later, the Chief Operating Officer was on a call with the company's largest customer, explaining why their order was delayed and what steps were being taken to prevent it from happening again.

This sequence of events is so common in organizations that it has earned its own economics. Management theorists call it the 1-10-100 Rule of Quality Costs: it costs one dollar to fix a problem at its source through prevention, ten dollars to fix it during production through correction, and one hundred dollars to fix it after it reaches the customer through failure recovery. That Connecticut motor was a one-dollar problem on Monday. By Thursday, it was a hundred-dollar crisis consuming executive attention.

The question that haunts modern operations is not whether organizations can detect problems early. They can. Sensors generate terabytes of data. Analytics platforms produce thousands of alerts. The question is whether the people closest to those problems have the information and the authority to solve them before they metastasize. In most organizations, the answer is no. And that gap, that hesitation between awareness and action, is where most operational value is destroyed.

The Hierarchy of Authority

The reason organizations struggle to delegate is not because frontline workers lack intelligence or judgment. The reason is risk. Senior managers accumulate decision making power because experience has taught them to foresee consequences that others might miss. They have watched small adjustments cascade into large disruptions. They have learned, often painfully, that well intentioned fixes can create new problems. The natural response is to centralize authority. If something important needs to change, it goes up the chain until it reaches someone with enough context to evaluate the tradeoffs.

This creates what organizational theorists call the Hierarchy of Authority: a structure in which the right to make decisions is determined by title. The manager approves because the manager's experience provides the judgment necessary to manage risk. The worker executes because the worker lacks that broader view. It is a logical system, and for much of industrial history, it worked well enough.

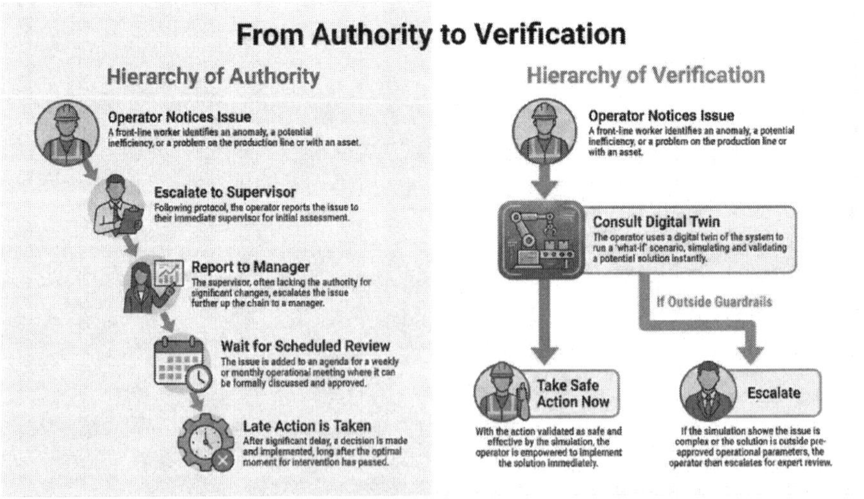

But the Hierarchy of Authority carries a hidden tax. Every decision that travels up the chain takes time. Every approval that requires a meeting delays action. Every supervisor pulled into daily troubleshooting is a supervisor not thinking about next quarter's capacity needs or next year's market opportunities. The Connecticut motor seized not because the organization lacked information, but because information alone does not confer authority. The worker knew something was wrong. The system did not give the worker the means to prove it was safe to act.

The Flight Simulator Effect

This is where the Digital Twin changes the organizational equation. Not by replacing human judgment, but by providing something that has historically been impossible: a risk-free environment in which frontline workers can test decisions before executing them in the real world.

Consider what happens when an airline trains pilots. No one hands a new pilot the controls of a commercial aircraft and says, "Let's see how you handle turbulence." Instead, the pilot spends hundreds of hours in a flight simulator, encountering every conceivable scenario without risking passengers or equipment. The simulator does not make decisions for the pilot. It gives the pilot the ability to make decisions with confidence because the consequences of error are contained within the simulation. When the pilot finally sits in a real cockpit, the decisions feel familiar. The judgment has been built. The risk has been managed in advance.

A Digital Twin offers the same capability to operational workers. An operator who notices a motor vibrating can consult the twin, which runs the data against its model of the system and simulates what happens next. Perhaps the twin responds: "Bearing failure is probable within 72 hours. Reducing speed by 5% extends operational life until the weekend maintenance window, with no impact on downstream production." The operator does not need a manager's permission to act on this information because the twin has already validated that the action is safe. The decision that once required escalation can now be made in minutes, at the source, by the person who first noticed the problem.

What this does, in effect, is replace the Hierarchy of Authority with what I call the Hierarchy of Verification. In the old model, you needed a title and an extensive authority matrix with complex delegation of authority to approve a decision because the organization depended on the manager's intuition to contain risk. In the new model, you need a valid simulation. If the twin verifies that an action is safe and effective, the worker has the confidence to solve the problem immediately rather than applying a temporary workaround and hoping someone higher in the chain eventually addresses the root cause.

Proof in Practice

Unilever discovered this when they deployed Digital Twins across their global factory network, from Brazil to China, for products ranging from Dove soap to Magnum ice cream. Their machines had always generated thousands of alerts daily. Under the old system, senior process engineers reviewed these alerts to determine when to stop a line for maintenance. It was a bottleneck that consumed expert attention and created delays.

With the Digital Twin in place, the diagnosis moved to the operator on the floor. If the twin predicted a motor issue, it alerted the operator directly, not the manager. The operator was empowered to adjust the machine's speed or schedule the maintenance window themselves because the twin had already calculated the safest option. The result was a 90% reduction in the number of alerts requiring management intervention. 90%. That is not an incremental improvement. That is a fundamental restructuring of how decisions flow through the organization.

Lowe's, the home improvement retailer, found a different application for the same principle. They created Digital Twins of their physical stores using NVIDIA Omniverse, fusing spatial data with inventory data in a real-time three-dimensional replica. Frontline associates were equipped with augmented reality headsets connected to this twin.

In the old model, associates who needed to check whether an item was in stock had two options: climb a ladder to physically inspect a high shelf or find a supervisor with system access. Both were slow. Both created dependencies. With the twin, an associate could look at a shelf and see through the cardboard boxes in augmented reality, instantly knowing what was inside without physical inspection or supervisory approval. Associates could see heat maps of customer traffic overlaid on the floor, identify dead zones where shoppers were not walking, and place digital markers to test new display layouts. Decisions that once required a regional merchandising planner could now be informed by the person standing in the aisle. The hierarchy of authority had been replaced by the hierarchy of verification. The twin validated the insight. The associate acted.

The Executive Dividend

There is a temptation to view this transformation purely through the lens of frontline empowerment, as if the primary beneficiaries are the workers who gain new authority. But the more consequential shift happens at the top of the organization.

When frontline workers can solve one-dollar problems at their source, those problems stop becoming ten-dollar corrections and hundred-dollar failures. The escalation ladder shortens. The approval chains thin. And executives find themselves with something they rarely have: time.

The Queen's Health Systems in Hawaii experienced this when they implemented a patient flow Digital Twin through their Aukahi Command Center. Hospital bed management had always been a source of constant escalation. When emergency departments backed up and intensive care units filled, decisions about discharges, transfers, and bed allocation bubbled up to senior administrators because no one at the operational level had a complete picture of the system's constraints.

The twin changed that dynamic. It modeled bed capacity and patient progression, predicting bottlenecks hours in advance. It flagged actionable blockers: discharge barriers, delayed imaging, bed turnaround delays. Critically, it gave charge nurses and bed managers the authority to make preapproved moves within defined guardrails. They could prioritize discharges, redirect housekeeping resources, and rebalance capacity across nursing units without waiting for executive sign off. If the predicted impact stayed below established thresholds, the operational team acted. Only when the impact threatened to breach those thresholds did the system escalate.

The results were striking. Average length of stay dropped by more than a day within ten months. Emergency department boarding fell by nearly

64%. Transfer acceptance rates increased by more than 22%. The hospital saved twenty million dollars annually. But the less visible outcome was equally important: senior administrators who had spent their days managing patient flow crises were now free to think about capacity expansion, strategic partnerships, and long-term operational investments. They had moved from being approvers of daily work to auditors of system performance. From firefighters to architects.

The Immune System

The biological metaphor here is useful. A healthy immune system does not wait for infections to become life threatening before responding. It detects pathogens at the cellular level and neutralizes them before they can multiply and spread. The body's senior management, the conscious mind, never even becomes aware of most threats because they are handled automatically by systems designed to respond at the source.

A Digital Twin creates an organizational immune system with the same properties. Small problems are detected and resolved at the frontline before they can compound into strategic disasters. The one-dollar issues stay one-dollar issues. The executives who once spent their Mondays investigating yield losses or their Fridays explaining supply chain failures to key customers can redirect that cognitive capacity toward growth, innovation, and competitive positioning.

This is the dividend that most organizations fail to quantify when evaluating Digital Twin investments. They model the operational savings: reduced downtime, improved throughput, lower defect rates. These are real and significant. But they miss the strategic value of executive attention. What is it worth for a CEO to spend two hours per week on market expansion instead of operational troubleshooting? What decisions does a COO make differently when freed from the daily approval queue? These benefits compound over time in ways that traditional ROI models struggle to capture.

The delegation dividend is real, but it is not automatic. Achieving it requires more than installing sensors and subscribing to software platforms. It requires understanding what a Digital Twin is actually made of: The layers of capability that must stack together to transform data into delegated decisions. Each layer serves a purpose. Each layer depends on the others. Skip any of them, and the hierarchy of authority remains firmly in place, decorated now with expensive technology that produces impressive visualizations but changes nothing fundamental about how decisions flow. That architecture is what we will examine next.

Chapter 5 – The Twin Stack

"The 'smart' layer fails when the 'boring' layers are missing"

Architecture of Value

A Digital Twin isn't a single technology. It's a stack of capabilities, each layer building on the one beneath it, each dependent on the integrity of what came before. Skipping a layer sharply increases failure risk. Some teams can approximate layers (e.g., manual action loops, partial truth feeds) and still produce value while maturing. But the most dangerous omissions are in the foundation: data plumbing and integration. Under-invest there, and even brilliant algorithms will produce confidently wrong outputs. Furthermore, if you under invest in the foundation of a Digital Twin (i.e., the lowest level), even the most sophisticated algorithms will produce meaningless results. Therefore, the understanding of this layered architecture will distinguish successful implementations from expensive failures.

Layer 1 – The Boundary

The first decision is not a technical decision; it is a conceptual decision: What is being twinned? Although this appears simple, it is not easy to define with precision. Are you creating a Digital Twin of a single piece of machinery, the entire manufacturing line or the entire manufacturing facility? Does the Digital Twin include the supply chain that brings raw materials into the facility as well as the distribution network that ships finished goods out of the facility? At which point does your system stop and the rest of the world begin?

This is important because it defines the scope of what you can optimize and what you must assume is outside of your control. For example, if your Digital Twin includes only the machinery, then upstream supply delays and downstream demand fluctuations are considered external events to which your system reacts. On the other hand, if your Digital Twin includes the entire value chain, then upstream supply delays and downstream demand fluctuations are now variables that your system can anticipate and potentially affect. There is no inherent reason that either choice is correct. In fact, making a poor choice or not making a conscious choice at all will ensure that your Digital Twin optimizes the wrong items.

Layer 2 – The Truth Feed

Once you have determined what falls inside your boundary, you need to know what is happening inside your boundary. This is referred to as the Truth Feed. I.e., the continuous feed of information that allows your Digital Model to stay synchronized with the physical environment. Sensors such as temperature sensors, pressure sensors, vibration sensors. Transactional information such as inventory movement, work orders, quality inspection transactions. Event information such as shifts changes, equipment states and environmental condition information.

In the physical environment, these are enabled by Internet of Things (IoT) devices. I.e., the connected sensors, cameras and smart equipment that are constantly streaming operationally related data. IoT devices act as the nervous system of the Digital Twin. IoT devices, use industrial protocols and edge computing platforms to provide the continuous data streams needed for the Digital Twin to be in sync with the real world.

This is where the unglamorous work resides: data plumbing, integrations, cleansing, tagging. Vendors rarely mention it in demonstrations. We will address this challenge in depth later. I have personally witnessed companies spend millions of dollars on Advanced Analytics Platforms and still have data infrastructure that looks like a collection of garden hoses duct-taped together. The algorithms were brilliant. The inputs were garbage. The outputs were confidently wrong.

The Truth Feed does not need to be perfect. "Perfection is the Enemy of Progress"; however, it needs to be honest about its imperfections. A Digital Twin that knows that it has uncertain inputs, can take those uncertainties into consideration. A Digital Twin that treats questionable inputs as gospel will make mistakes with absolute confidence.

Layer 3 – The Model

The Third Layer is where your Digital Twin obtains its knowledge of how things work. This is the Model Layer: The mathematical and logical representations that define the behavior of your system. Here you must make a fundamental choice between two philosophical approaches.

1. **Physics-Based Models** which codify the laws that govern your process. They know that heat transfers according to thermodynamic equations, fluids flow according to pressure differentials, mechanical stress accumulates according to material properties. physics based models are interpretable and stable. They will not suddenly predict that metal melts at room temperature. However, building these models requires deep domain expertise and may miss subtle patterns that do not fit neatly into a known equation.
2. **Data-Driven Models** which are typically machine learning that learn directly from historical observations. They do not need to understand why a pump fails; they only need enough examples of pumps failing to recognize the precursors of failure. Data Driven Models can capture complex, non-linear relationships that physics alone may miss; however, they are only as good as the training data they have seen. When asked to extrapolate beyond their training experience, they can produce nonsense with no warning.

The best Twins use both approaches. Physics to establish guard rails and machine learning to identify nuances within those guard rails. We will return to hybrid approach when we discuss governance, as it is one of the most effective tools to prevent digital hallucinations.

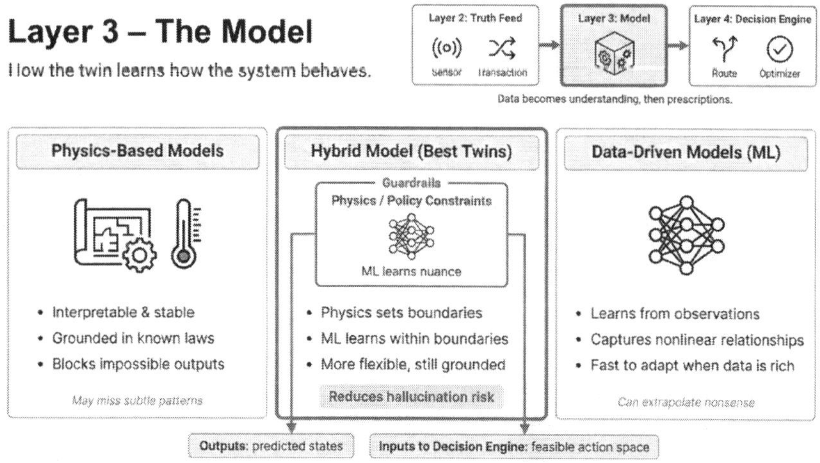

Layer 4 – The Decision Engine

The fourth layer converts understanding into action. This is the decision engine. The forecasting and optimization logic that translates model outputs into prescriptions. It is the layer that answers the level 4 question from our maturity ladder "What Should Be Done?"

The decision engine must hold multiple considerations in tension simultaneously. It knows what the model predicts will happen and what outcomes you want to achieve. It knows what constraints you cannot violate e.g., safety limits, contractual commitments, resource availability. It searches the space of possible actions to find the prescription that best balances all these factors. Not the theoretically optimal solution, but the practically executable solution.

Layer 5 – The Action Loop

The Fifth and final layer is where the digital world reconnects to the physical. The Action Loop: The mechanism through which the twin's recommendations change reality. Be it through a human who receives a recommendation and executes it or through a system that acts automatically. Additionally, this layer closes the feedback circuit: Actions create outcomes, outcomes generate new signals, and those signals flow back into the truth feed to update the twin's understanding.

The choice between human-in-the-loop and automated execution is not a binary one. Most mature twins operate along a spectrum: Routine decisions made within established parameters execute automatically while novel situations, edge cases and high-stakes choices escalate to human judgment. The key is designing this spectrum intentionally rather than discovering it accidentally when something goes wrong.

By now, you understand what a Digital Twin consists of. The next logical question is where should you build a Digital Twin? Not every operational challenge is suitable to being addressed using this methodology and not every DT produces equal value. In the next chapter, we will examine the conditions that separate high-leverage opportunities from expensive distractions and identify the two starting points that always provide the fastest path to measurable impact.

The Twin Stack

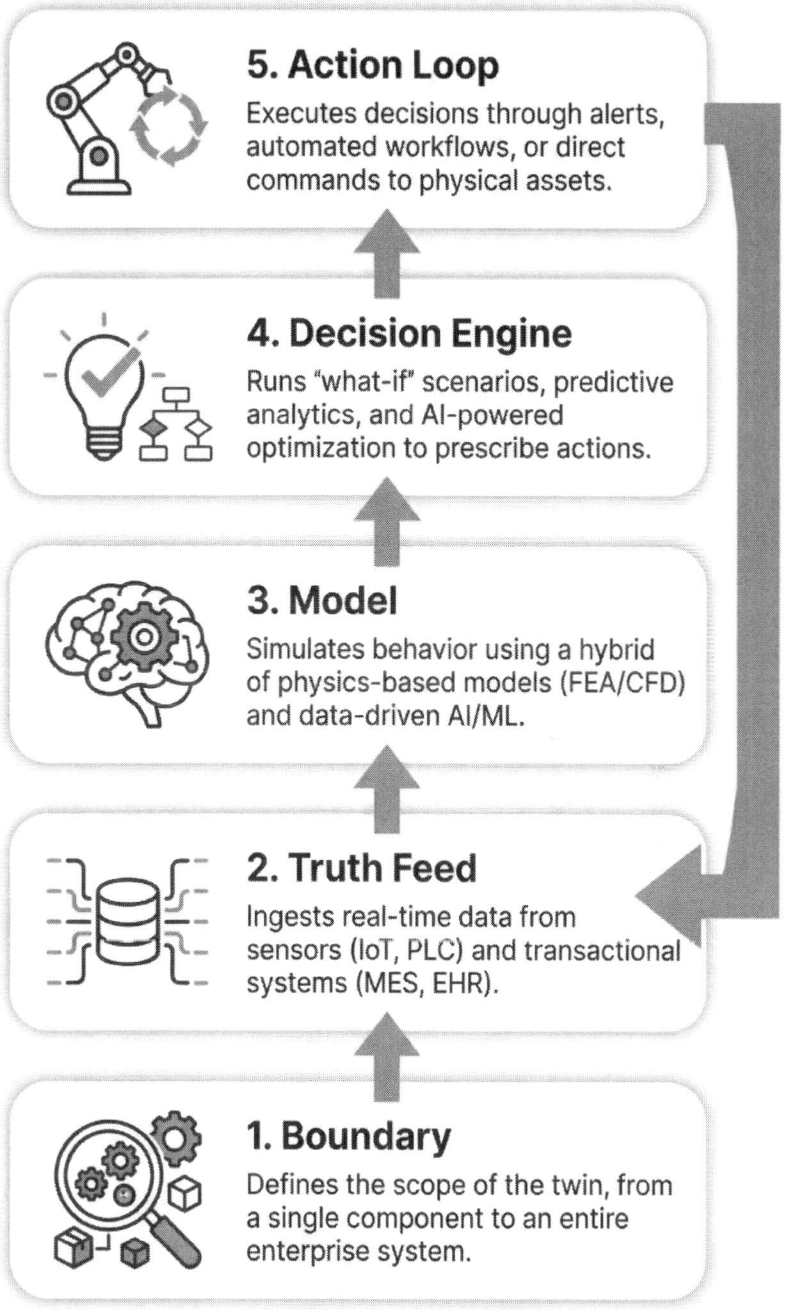

Chapter 6 – Identifying the Opportunity

"Start where decisions repeat and variance is expensive. Everything else is theater."□

Not every problem needs a Digital Twin. That may seem like an odd thing to say after writing five chapters about Digital Twins, but it is perhaps the most critical strategic insight I can give you. The companies that do poorly with DTs tend to fail for the same reason: they create a DT where a DT is not required, or they created a DT to address problems which cannot be resolved with a DT. The hammer was a great hammer. It was simply the wrong tool for the job.

The issue is not whether DTs are powerful, they are. The question is whether your operational problem has characteristics which make a Digital Twin the best solution to apply. If you get this wrong, you will invest months developing complex infrastructure to solve an operational problem which could have been solved using a spreadsheet and a regular meeting.

The Three Conditions Litmus Test

Repeated implementations across industries and sectors., three conditions reliably separate high-leverage twins from expensive distractions. All three should be present before a twin becomes a necessity.

1. **Repeatable Decisions**: Repeatable Decisions: Digital Twins provide the greatest value when identical types of decisions are made frequently. For example, staffing for the next shift. Determination of delivery routes for the next 100 customers. Scheduling maintenance for a group of machines. While each individual decision may be trivial, the collective impact of each decision can be substantial and the consistency of optimizing each decision

will compound over time far greater than the impact of an occasional brilliant human decision.

2. **Fixed Constraints:** Digital Twins operate optimally when the environment is governed by laws of physics, regulatory requirements, contractual commitments, etc., that cannot be negotiated away. A jet engine operates under thermodynamic limits. Hospitals are obligated to meet nurse/patient ratio standards. Chemical processes must be conducted within certain temperature ranges or risk catastrophic failure. The fixed constraints define the solution space within which the Digital Twin will search for optimization solutions. The Digital Twin does not have to consider limitless alternatives. Instead, it should seeks solutions within the constraints established by the laws of nature, regulation, contract and/or other non-negotiable boundaries.
3. **High Cost of Variance**: Digital twins will produce the most value when the differences between good decisions and bad decisions are measured in significant financial, safety related, or customer satisfaction terms. Downtime on a production line those costs $50K per hour. A stockout causing loss of a major customer. A delayed diagnosis that prolongs a hospital stay by 3 days. High cost of variance creates a scenario in which

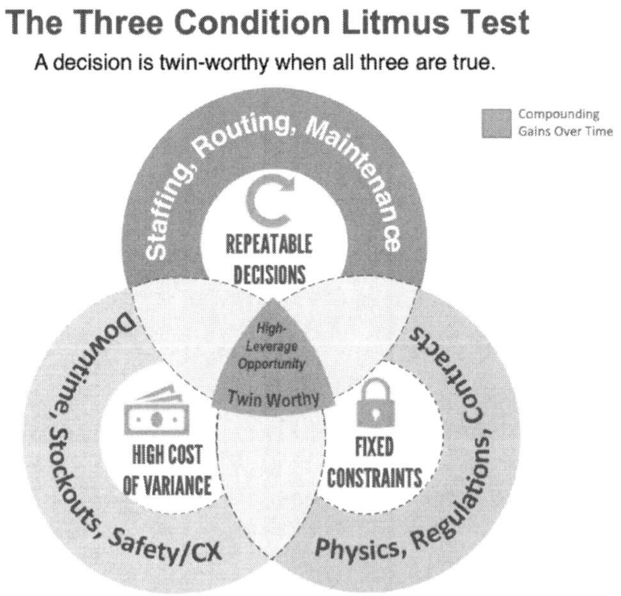

the Digital Twin's ability to consistently identify superior answers provides a rapid return on investment. Low cost of variance usually results in little justification for the investment.

Where Twins Don't Belong

Understanding how Digital Twins succeed requires understanding where they fail. Creativity and innovation are resistant to DTs since there are no constraints to optimize against and there are no objective functions to maximize. Digital twins are usually the wrong tool for open-ended creative work such as marketing copy, brand voice, product aesthetics, because objectives are difficult to formalize and "better" is largely subjective. Generative AI fits those problems more naturally. Twins excel where constraints exist, outcomes are measurable, and decisions repeat.

This means that Digital Twins are developed to solve problems that have a scoreboard. An event has occurred; an event has not occurred. A production line was down; a shipment failed to meet the window. A patient had to wait longer than necessary. Because events are quantifiable, a twin can learn, and improve; a twin can develop trust.

Therefore, twins are less useful when "better" is largely a function of taste or politics. A board may debate whether a new corporate branding is modern; there is no science to reference, and no real-time feedback mechanism that would report back to the board within a week or month to provide a definitive answer. These types of environments (where the answer is subjective), generative AI can assist in gathering examples and in developing supporting arguments; however, a Digital Twin should never be sold as the decision engine.

Where the problem is a system that is repetitive, defined, and costly to make mistakes on, this is where the twin fits.

The Power Pair

For executives seeking to initiate a Digital Twin program, I recommend viewing opportunities through two value lenses that nearly always survive their first encounter with reality. Both lenses focus attention on decisions that are repetitive, constrained, expensive to err on, and quantifiable. They connect directly to the income statement. I refer to these two perspectives as the Power Pair: the Cost Out and Productivity Twin (CPT) and the Growth and Revenue Twin (GRT). One lens focuses on reducing costs and variability: the other focuses on converting operational certainty into revenue. Most successful Digital Twin initiatives begin by examining their operations through one or both lenses to identify where a twin will generate the fastest, most defensible return.

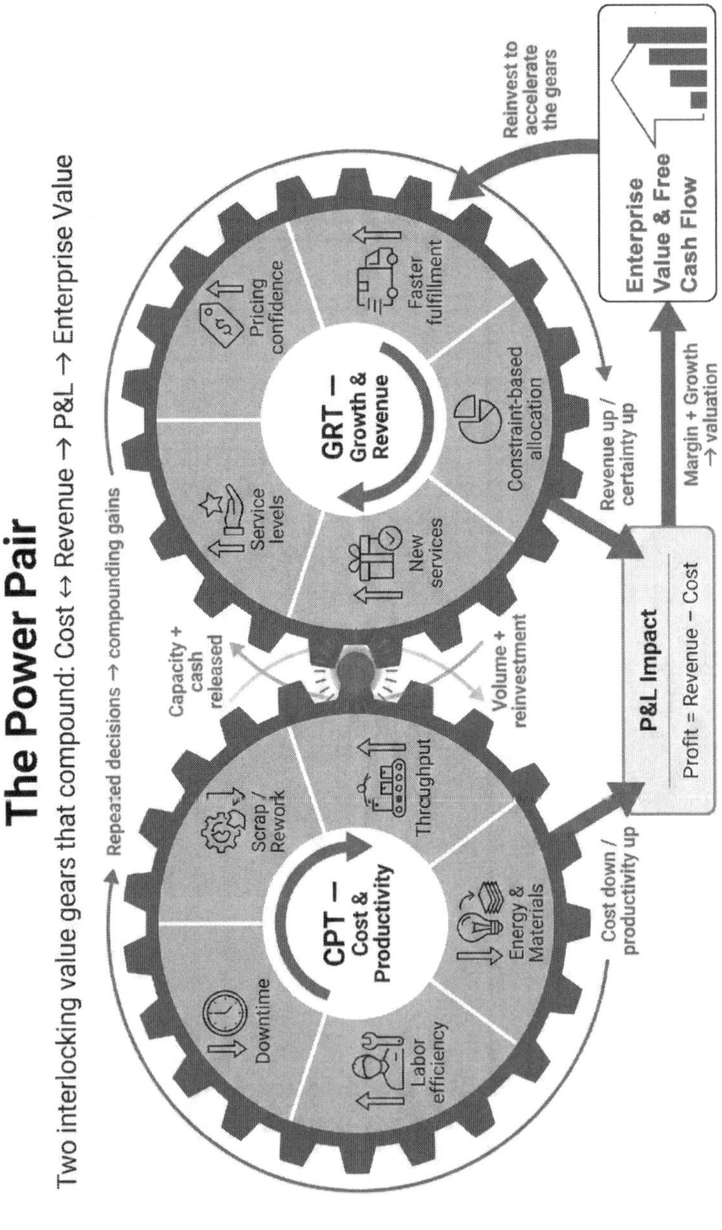

The Cost Out and Productivity Twin (CPT)

The Cost Out and Productivity lens focuses on eliminating unnecessary spending in operations: downtime, scrap, wasted energy, emergency shipping, overtime, and the silent tax of fighting fires. Predictive Maintenance is typically one of the first implementations viewed through this lens, because predictive maintenance is quantifiable. Assets emit signals before they fail, and failing is expensive.

A CPT maintains a live representation of the health of the assets and the constraints on the operation. It predicts when the asset is likely to fail, identifies the last responsible moment to take action, and schedules the required work to allow the plant, fleet, or facility to absorb it. Over time, the same loop can be repeated to apply to other areas of preventable waste that exhibit similar properties: yield loss, energy peaks, and all of the small decisions that generate large amounts of waste.

The CPT lens produces a double benefit from operational variability: it both reduces avoidable costs and increases output per asset. CPT models provide an active, living view of the operation on the shop floor, in the warehouse, or in the field, which continuously synchronizes to the actual operation. In addition, CPT models recommend the next best action based on the current, real-world constraints.

In most cases, twins built through the CPT lens will focus on a few tightly defined goals:

- **Reliability**: reduce unplanned downtime, increase uptime, decrease repair time
- **Quality and Yield**: reduce scrap and rework, increase first pass yield
- **Energy and Materials**: decrease energy used per unit, reduce material usage
- **Labor Productivity**: reduce overtime and expediting, improve equipment use
- **Flow and Throughput**: maximize production through the constraint, reduce change over losses, stabilize schedules

Predictive Maintenance is a typical entry point for CPT initiatives, as the financial impact is immediate. The CPT model estimates the likelihood of equipment failure and recommends the optimal time to take corrective action: either immediately, at the next available safe opportunity, or continue running and risk additional loss due to equipment failure. Predictive Maintenance eliminates a common dilemma: maintaining equipment too soon results in wasted capacity; maintaining too late results in lost production, scrapped goods, and delayed shipments.

Throughput gains are typically achieved by protecting the constraint, not working harder. The CPT model identifies the true constraint, evaluates sequences and schedules based upon potential disruptions, and adjusts the schedule when disruptions occur (delays in receiving raw materials, wear and tear on machinery, or absenteeism) to ensure that production continues without interruption.

Quality, energy, and waste act as quiet multipliers. Quality oriented twins connect process condition variables to the likelihood of defects occurring and provide recommendations to adjust those variables to minimize the risk of producing defective parts. Energy oriented twins optimize process parameter settings relative to efficiency curves and production priorities. Even small percentage improvements can result in sustainable margins.

The litmus test is simple: the twin must change the work. Recommendations made by the CPT model should be actionable and directed to specific areas within the execution systems (work orders, scheduling, dispatch, operator guidance), with well defined guardrails, escalation rules, and measurable outcomes. When this loop is closed, cost reduction becomes continuous, and productivity gains are sustained over time.

The Growth and Revenue Twin (GRT)

If the Cost Out and Productivity lens is about removing friction from operations, the Growth and Revenue lens is about something more ambitious: converting operational certainty into commercial advantage—whether by creating entirely new revenue streams, transforming pricing models, or extracting value that was previously invisible.

The "as a Service" (aaS) business model in the logistics and transportation sector is a prime example of how a Digital Twin can be adopted to create new revenue streams. Michelin has transformed portions of its commercial fleet business from selling tires as hardware to selling "Tires as a Service" through its Fleet Solutions division. In this model, the fleet operator does not buy the tires. Instead, Michelin retains ownership of the tires, installs them on the fleet's vehicles, and bills the customer based on kilometers driven.

In this relationship, the fleet operator wants maximum fuel efficiency and zero uptime loss due to tire failures. Michelin wants to maximize the lifespan of the tire and ensure the casings can be retreaded. The Digital Twin enables this alignment of interests. By embedding sensors (TPMS and RFID) inside the tires, Michelin creates a Digital Twin of the rolling fleet. This model continuously monitors pressure, temperature, and wear in real-time, analyzing how the tire interacts with the specific routes and loads of that customer.

Because the twin can predict wear rates and detect pressure anomalies before a blowout occurs, Michelin can proactively schedule maintenance and rotate tires at the exact moment required to extend their life. This allows Michelin to assume the risk associated with owning the asset. They can confidently offer performance guarantees—such as specific fuel savings targets—because the twin's predictions allow them to intervene before problems erode those guarantees. The customer pays a premium for the certainty of uptime and fuel efficiency, and Michelin extracts value from the asset's full lifecycle that was previously lost when they simply sold the hardware and walked away.

The evolution of Rolls-Royce's "Power by the Hour" model illustrates how Digital Twins transform the economics of outcome-based pricing. The business model itself is not new. In 1962, Bristol Siddeley introduced a fixed-cost-per-flying-hour maintenance program for the Viper engine, which powered the de Havilland 125 business jet. The Viper had been designed for short-life military drones and required frequent maintenance when adapted for corporate aviation. To reassure customers, the manufacturer offered complete replacement and maintenance coverage for a predictable hourly fee. Rolls-Royce acquired Bristol Siddeley in 1966, inherited the program, and eventually trademarked the name "Power by the Hour."

For decades, this model operated on actuarial faith. The manufacturer priced the contract based on historical failure rates and average maintenance costs, then hoped reality would cooperate. The economics worked in aggregate, but individual engines could produce painful surprises. Every unplanned failure ate into margins. Every conservative maintenance schedule left money on the table.

The Digital Twin changed the underlying math. Today's TotalCare program—Rolls-Royce's modern incarnation of Power by the Hour—uses real-time IoT sensors to monitor engine health continuously during flight. The twin tracks temperature profiles, vibration signatures, pressure differentials, and hundreds of other parameters, comparing actual behavior against physics-based models of expected degradation. When anomalies emerge, the system predicts failure windows and schedules intervention during periods that minimize customer disruption.

This is where the strategic shift occurs. The original Power by the Hour transferred risk from customer to manufacturer through contractual structure. The Digital Twin transforms that transferred risk into manageable, optimized operations. Rolls-Royce no longer prices contracts on historical averages and hopes for the best; it prices them on real-time visibility into the actual condition of each specific engine. The manufacturer can now offer outcome guarantees with confidence because it can see problems developing and intervene before they become expensive. The customer pays a premium for certainty. Rolls-Royce captures margin that would have been impossible when maintenance decisions were based on calendars and flight-hour thresholds rather than actual component health.

The lesson is not that Digital Twins invented outcome-based pricing—the business model predates the technology by decades. The lesson is that Digital Twins make outcome-based models economically superior for the provider. What was once a calculated bet becomes a controlled system. What was once risk transfer becomes risk elimination.

This pattern is spreading. Kaeser Kompressoren, a German manufacturer of industrial compressors, now sells cubic meters of compressed air rather than compressor hardware to many customers. The customer only pays for what they actually consume, and that consumption is measured and verified using the Digital Twin. Kaeser maintains the compressor equipment, ensures that the equipment operates at peak efficiency, and extracts the gains associated with increased efficiency that were previously unknown. The Digital Twin functions as both the operational intelligence that ensures the system continues to operate properly, and as the verification mechanism that proves to the customer what value they have received. If the twin did not provide the second function, the trust required to establish an outcome-based contract would never exist.

That said, the GRT lens applies far beyond servitization models. Many of the most advanced applications of GRT are examples of what could be described as dynamic value extraction: the capability to understand individual customer willingness to pay and then adjust offers accordingly.

Banks and retailers have long used demographic information and purchase histories to segment their customers. Digital twins take this capability to a different level entirely. Rather than placing customers into large groupings based on demographics and purchase history, advanced commercial systems are now developing personalized behavioral models that combine propensity scoring, uplift modeling, and in some cases reinforcement learning to predict how a specific customer will respond to different price points, promotional offers, and product configurations. The system tests multiple options against the individual customer model prior to providing the option most likely to result in conversion at the greatest profit margin. This is not about applying the highest price possible to every customer. It is about recognizing that some customers are price-sensitive and need a discount in order to induce conversion, while others would have purchased at the standard price anyway. The twin identifies which customer is which and adjusts accordingly. The same reasoning applies to retention. If the behavioral model indicates that a customer is likely to leave, the system can automatically initiate a retention offer that is tailored to the predicted price-sensitivity of that specific customer.

The manufacturing world has discovered its own version of this dynamic: the virtual sensor. Physical sensors are expensive to install and maintain, particularly in harsh operating environments. But a well-calibrated Digital Twin can infer data that would otherwise require hardware. Consider an engine manufacturer that wants to offer customers insight into internal combustion temperatures. Installing a physical sensor in that location might be impractical or prohibitively expensive. Instead, the twin uses data from other sensors combined with physics-based models to calculate what the temperature must be. The customer receives the information they need without the hardware cost. The commercial implication: the manufacturer can offer this capability as a premium subscription. The base machine comes with standard instrumentation. If you want the virtual temperature reading, you pay extra. The customer is paying for a data point generated entirely by an algorithm, effectively turning the twin into a product feature that can be sold alongside the physical equipment.

The same logic applies to what might be called the transparency premium. Many industrial customers would pay for access to a dashboard that visualizes the hidden internal state of their assets: stress accumulation in critical components, efficiency degradation over time, remaining useful life estimates. This information has always existed, locked inside the physics of the machine. The twin makes it visible and therefore sellable. What was once a one-time hardware transaction becomes recurring software revenue.

Supply chain twins unlock an entirely different category of revenue opportunity: yield management through demand-constraint visibility. When a supply chain twin can visualize global inventory and production capacity in real time, predicting bottlenecks weeks before they materialize, the commercial implications are profound. A manufacturer facing future production constraints can identify which orders to accept, which customers to prioritize, and which delivery slots to hold for premium pricing. When the twin predicts scarcity, the sales system can flag future production slots as premium, charging expedite fees that reflect actual constraint economics rather than arbitrary markup tables—or, in theory, allowing priority customers to bid for scarce slots. Few organizations have implemented true auction mechanisms for production capacity, but the enabling capability—real-time constraint visibility—now exists. The same visibility enables what might be called intelligent inventory rationing. If the twin predicts a demand spike in two weeks due to seasonal patterns or market conditions, it can reserve current inventory for the high-value customers who will emerge during that spike rather than depleting stock through routine orders today.

Energy markets have pushed these concepts further still. A Digital Twin of a battery storage system simulates both the battery's health characteristics and external market prices. The optimization engine holds energy when prices are normal and releases it only during peak windows when prices spike. This is pure arbitrage: buying low, selling high, but executed with a precision that requires understanding both the market dynamics and the physical constraints of the asset. The twin ensures that aggressive discharge patterns do not damage the battery, balancing short-term revenue capture against long-term asset health.

Perhaps the most unexpected revenue stream is one that creates value from simulation rather than observation: synthetic data. Every simulation run by a Digital Twin generates information about scenarios that never actually occurred. A hospital system that runs millions of patient flow simulations creates data about disease progression, treatment outcomes, and resource utilization patterns that would be impossible to collect through observation alone. Real-world data is messy, expensive to gather, and often restricted by privacy regulations. Synthetic data generated by a well-calibrated twin is clean, unlimited in quantity, and significantly reduces re-identification risk—though it requires validation to ensure the underlying distributions do not inadvertently leak information about real individuals. When properly tested, synthetic data offers a path to training models on scenarios that would otherwise be impossible to observe. Healthcare systems, city planners, and industrial operators are beginning to recognize that this synthetic data represents an asset that can be sold to third-party researchers, pharmaceutical companies, technology firms, and academic institutions. The twin is not just informing internal decisions; it is manufacturing a product from algorithmic imagination.

What unites these implementations is the use of operational certainty as commercial leverage. The mechanisms differ—outcome-based pricing, yield management, arbitrage, data monetization—but the underlying capability is the same: a twin that knows the state of the system well enough to price, commit, and act with confidence competitors cannot match. In each case, the twin reveals value that was previously invisible: customer willingness to pay that went undetected, operational certainty that could not be monetized, constraint economics that were averaged away in standard pricing, data exhaust that was discarded rather than sold.

Twins built through the Growth and Revenue lens do not replace traditional commercial functions. Sales teams still sell. Marketing still generates demand. Pricing committees still set strategy. But the twin changes the information environment in which all of these functions operate. Decisions that were once made on intuition and averages can now be made on individual customer models and real-time constraint visibility. The result is not just more revenue but higher-quality revenue: deals with better margins, customers with higher lifetime value, and capacity allocated to uses that maximize return.

These two value lenses represent the ideal starting point for evaluating Digital Twin opportunities because they address both sides of the income statement. The CPT lens identifies where twins can reduce costs and variability by removing preventable waste and failures. The GRT lens identifies where twins can increase revenue by improving the quality of commercial decisions: what to commit to, to whom, at what price, and with what level of confidence. Used together, they reveal opportunities that create a business case even the most skeptical CFOs can audit.

Frameworks and logic can only take things so far. At some point, executives will want to see empirical evidence that these ideas can be successfully applied in practice using actual data from actual organizations experiencing actual operational challenges. We will review a number of case studies from manufacturing, healthcare, and infrastructure in the next chapter, and demonstrate what can be accomplished when the Mirror Strategy is applied effectively.

Chapter 7 – Proof in the Numbers

"If value is not measured by design, it will be debated by accident."

The frameworks discussed in previous chapters guide selection. E.g., where to look, what conditions to test, which value lens to apply. But selection is not proof. At some point, every executive facing a Digital Twin investment needs to answer a simple question: does this actually work? Not in theory, not in vendor demonstrations, but in real operations with real constraints and real money at stake. The answer, as it turns out, is yes. But the details of how it works are more instructive than the fact that it does.

The Holistic Value Play

Most people use Unilever's products daily. Think soap, shampoo, ice cream, mayonnaise, but rarely consider what is involved in the complex process of creating each product. In addition to the chemistry required for each product type, Unilever also had to account for the different types of equipment used and the various ways those pieces of equipment can fail when they do fail. When the company decided to implement Digital Twin technology at eight of its facilities, the natural assumption would have been to test the new digital tools first in a single plant and if successful, roll out to all remaining plants (the "test a little, learn a lot" model). However, Unilever approached this project with a unique, total factory vision and implemented Digital Twins that linked equipment health to scheduling to quality output across all eight of its facilities.

The financial impact was something Unilever's finance team could quantify directly: $52 million in documented savings over the initial implementation period, driven by improved asset utilization and reduced reactive maintenance across the eight facilities. While these are great metrics, the real benefit of the Digital Twin program lies in Unilever's ability to better understand how problems develop and grow as a result of the connection between its various processes and equipment. For example, a compressor malfunction in one area of a plant does not only cause problems for the compressor itself, it causes quality issues downstream, scheduling changes due to shift changes, and inventory misalignments that continue on to other areas of the facility and potentially other plants. Therefore, the ability of the Digital Twin to demonstrate and track these interconnects changed maintenance from being a cost center to a strategic asset for the company.

The Closed Loop

In contrast, General Motors has another problem to contend with. In the case of forming auto body panels using massive, high tolerance stamping presses, failure of the presses causes an entire production line to come to a halt. Employees are then left standing idle as vehicles cannot be produced and therefore customers will not receive their vehicles in a timely manner. Unplanned downtime costs for GM in this area can be quantified as maintenance expense plus lost production time, and expedited shipping and loss of customer relationship damage.

The Digital Twin project at GM was directed towards improving performance in stamping operations and the resulting implementation created what engineers refer to as a "closed-loop" process. Data from sensors installed in the presses is fed into the Digital Twin, which continuously monitors and compares current signature data from the presses to historical baseline data to identify any anomalies that may indicate impending failure. If the Digital Twin identifies any anomalies, it does not merely alert maintenance personnel, but instead provides specific recommendations for intervention, scheduling those recommended interventions during scheduled maintenance windows when possible, to minimize the impact on production.

As a direct result of implementing the Digital Twin, General Motors experienced a 25% decrease in unplanned downtime for its stamping operations. However, the more compelling results were behavioral changes within the organization. For years, maintenance technicians had been forced into reactive firefighting. After the twin was implemented, the consistent accuracy of its prescriptions gradually earned technician trust, allowing them to shift from responding to emergencies to preventing them. The twin did not replace technician expertise; it amplified it by surfacing patterns that humans recognize less reliably, freeing technicians to focus on the mechanical judgment that machines cannot replicate.

Unlocking Latent Capacity

Hospitals are one of the few places where you will see the equipment being people and the product being their health and wellness. Queen's Health Systems (Hawaii) had a problem similar to many hospital directors. They were experiencing growth in demand but were limited in the number of beds available. There was also a lack of funding to build additional capacity.

In the past, hospital administrators have responded to this issue by building additional capacity. However, this process can take years and cost hundreds of millions of dollars. Therefore, what if you could provide additional capacity without having to lay a single brick?

Queen's created a Digital Twin of the patient flow at Queen's Hospital in order to model how patients moved through the healthcare delivery system from admission to discharge. This Digital Twin tracked such items as bed occupancy, scheduling of procedures, test results, discharge status, etc., and dozens of other variables which determined how long patients remained in the healthcare system. As well, it identified bottlenecked areas within the healthcare delivery system which could not be seen by any one individual department due to interaction between departments. A delay in obtaining lab results did not only cause delays in diagnosis, however, it caused delays in procedures, which caused delays in discharging patients, which caused delays in admitting new patients, which resulted in diverting ambulances to other hospitals.

By improving patient flow, Queen's reduced average length of stay by 0.07 days per admission. That number sounds trivial until you run the math. Across 150,000 annual admissions, 0.07 days translates to approximately 10,500 freed bed-days per year. At an average daily bed cost of $2,000, this represents $21 million in unlocked capacity annually—the functional equivalent of building a 30-bed wing without laying a single brick or hiring a single additional nurse. The twin did not add resources; it removed friction that was invisible to any single department. Overall, patients received care more quickly. The staff of the hospital experienced less chaos. The overall financial model of the hospital improved significantly.

Detecting the Hidden

Thames Water provides water services to 15 million customers in and around London using a 100-year-old plus network of pipes, and as such loses water due to leaks. While water loss from pipes is unavoidable as part of the aging process of infrastructure, the scale of the issue is extremely high. Before their Digital Twin solution was implemented, Thames Water was experiencing water loss at levels that were financially damaging and environmentally unsustainable.

Leaks in a network are usually hidden and therefore very difficult to detect and quantify. Although water leaking into the ground will eventually show up on the surface when a pipe bursts, the leak may have been causing water loss for months before appearing. Thames Water's Digital Twin uses a large number of IoT sensors which are located all over its distribution systems including flow meters, pressure sensors and acoustic monitors, which feed continuous streams of data to the central model. These IoT sensors allow for comparisons between the expected (modeled) flows as opposed to the actual (measured) flows in each part of the network. The Digital Twin model allowed Thames Water to create a digital representation of its water distribution network and compare modeled expected flows to actual measured flows throughout the network. As a result, if there was a difference (or 'discrepancy') between the modeled expected flow and the actual measured flow, it would indicate there had been a hidden loss of water somewhere within the network. The Digital Twin model would provide information about not only that there was a loss of water, but also approximately where in the network the loss was occurring.

The results of this approach were 18 million liters of water saved each day. This means that the water did not need to go through treatment, pumping and delivery processes. In addition, the water loss resulted in reduced environmental impacts and reduced operational costs. The Digital Twin did not repair or replace the damaged or failing pipes; it identified where the damaged or failing pipes existed so that maintenance teams could focus on repairing those pipes. In essence, an insurmountable problem of inspecting thousands of kilometers of pipeline, was transformed into a much smaller, focused inspection effort to address the highest impact issues first.

Calculating The Business Case

The cases presented here have one key factor that has led to justification for their respective investments. That factor is either avoiding pain or unlocking an existing constraint. The "avoiding pain" refers to preventing costly negative events (e.g., downtime at GM; detection of leaks at Thames); the "unlocking constraints" is the ability to derive additional value from existing assets (Queen's increased capacity; Unilever's increased throughput).

When building your business case, identify where your twin will create value (either avoiding pain or unlocking a constraint) and provide detailed numeric information regarding the value your twin will generate. The positive results of these cases did not occur overnight. All four organizations – Unilever, General Motors, Queen's Health Systems, and Thames Water – pursued a methodical process of implementing their twinned technology and each organization was able to avoid a very common pitfall when it came to their twin implementation: attempting to "boil the ocean."

They all implemented the initial phase of their twinned technology with limited scope, were able to demonstrate immediate value, and then expand their twinned technology based on successful implementations. We will outline this methodology in detail in the next chapter titled "Minimum Viable Twin," that you may apply to your own twin implementation.

Chapter 8 – The Minimum Viable Twin (MVT)

"Speed comes from cutting scope, not cutting truth."

In my experience, seven of ten failed Digital Twin programs began with enterprise-wide scope rather than focused pilots. That pattern appears counter intuitive. We celebrate ambition, applaud bold vision, and tell ourselves that thinking big is how transformation occurs. We celebrate ambition. We applaud bold vision. We say to ourselves that thinking big is how transformation occurs. However, in Operational Technology, an abundance of ambition without discipline produces expensive disappointments. Companies almost always create successful Digital Twins by starting small enough to feel uncomfortable, proving value quickly, and then expanding once the company has earned the right to do so.

I have seen far too many companies initiate eighteen-month programs to develop comprehensive Digital Twins of their total operations only to have those programs fail at approximately month 14 due to an explosion in programmatic requirements, difficult-to-complete integrations, and frustrated business sponsors unable to wait another six months for value. Millions of dollars had been spent, and in the end, all that was left was a warning story and an organization unwilling to try again.

There is an alternative called the Minimum Viable Twin (MVT), which is an implementation strategy that includes a focused implementation that can be developed in thirty to sixty days, demonstrates value on a single decision, and establishes the platform for disciplined expansion. It is not a compromise or a half-measure; it is a strategy.

The 30-60 Day Strategy Playbook

The first task may be the most important, and it is the area where most teams go wrong immediately. You must select one decision, not one department, not one process, nor one asset class. One decision that is repeated, and where the improvements are quantifiable. Not "Optimize Maintenance," but "Determine when to replace the seals on Pump 7." Not "Improve Patient Flow," but "Decide Which Patients Are Ready For Discharge This Morning." The narrower the focus, the quicker you will know if the strategy is working.

To executives who have been educated to think about transforming the Enterprise, choosing a single decision feels wrong. Is the value of a Digital Twin realized through viewing the whole system? Yes, eventually. However, you cannot view the whole system until you demonstrate you can clearly view one decision. The error is equating the final destination to the first step.

The second step is to define the boundaries and the guardrails for the twin before developing the twin. What boundaries will the twin not cross? What safety limits are non-negotiable? What regulatory standards cannot be deviated from? What business rules represent contractual or policy obligations? The boundaries and guardrails represent the foundation of the organization's trust in the twin. The twin is more likely to gain increased levels of autonomous operation as long as the twin remains within the established boundaries and guardrails. A twin that proposes impossible or hazardous decisions will never establish trust regardless of the sophistication of its algorithms.

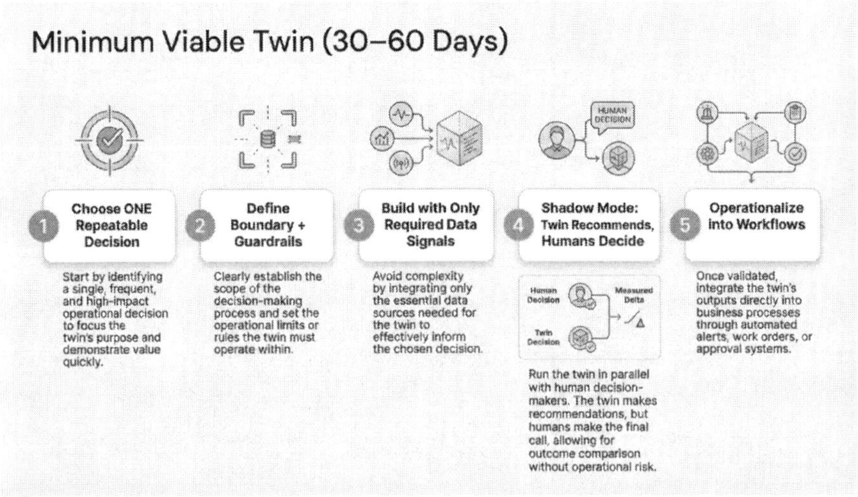

The third step is to build the MVT itself, and at this point, discipline is necessary. You only require the data needed to make the one decision.

All the data you plan to use later. The comprehensive data warehouse your IT team has been planning for three years. Only the data signals required to answer today's question. If you are determining when to replace the seals on a pump, you need vibration data, run-time hours, and failure history for that pump type. You do not need the entire maintenance management system to integrate end to end. The most common source of scope creep during this phase is why thirty-day projects become eighteen-month disasters.

This is where discipline matters most. An MVT does not solve the enterprise data infrastructure problem. It bypasses it temporarily to prove value. In practice, this often means working with manual data exports, flat files, or direct API access to a single source system rather than waiting for the data lake integration that IT has been planning for three years. The goal is to demonstrate that the decision logic works before investing in the plumbing that scales it. Once the twin proves its value on constrained data, the business case for infrastructure investment writes itself.

Learning in the Shadows

The fourth step represents the majority of the learning: operating in shadow mode. Shadow mode involves the twin providing recommendations to humans, who continue to make decisions as they normally would. Both sets of decisions are documented and compared. Were the twin's recommendations consistent with what the human selected? Where did they differ, and what was correct? What patterns emerge in the differences?

Shadow mode serves several functions simultaneously. It verifies whether the twin's prescription is truly superior to current practices. It develops trust with the operators who will ultimately execute on those prescriptions. It identifies areas that represent the twin's "edge cases," where the model requires further refinement. And finally, shadow mode produces the performance data you need to justify scaling. A twin that consistently demonstrates superiority over human decision-making in shadow mode has demonstrated that it has the right to enter into production. A twin that does not produce such results has indicated issues that can be resolved prior to the costs being incurred.

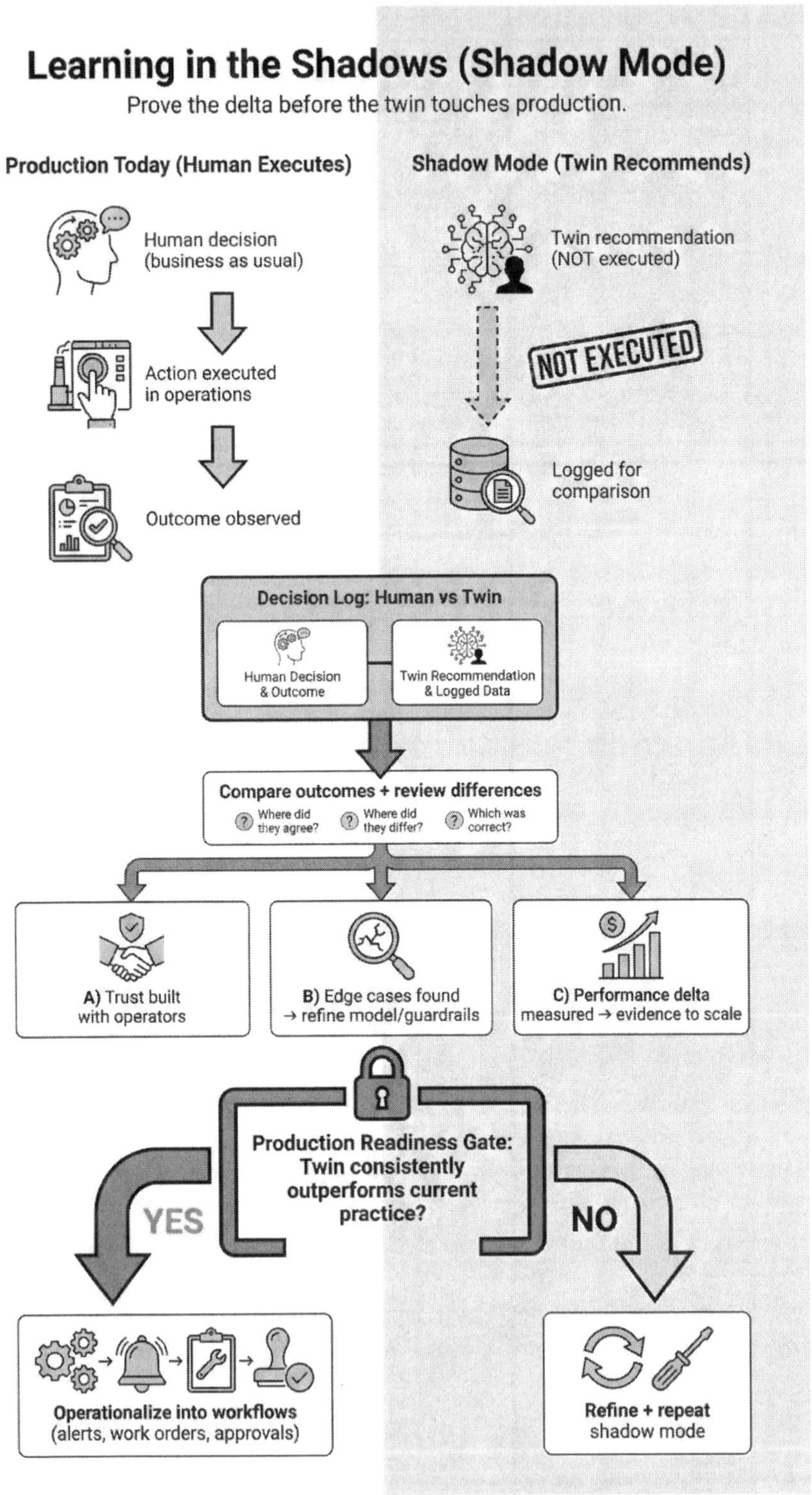

Learning in the Shadows (Shadow Mode)
Prove the delta before the twin touches production.
Production Today (Human Executes)
Shadow Mode (Twin Recommends)
Human decision (business as usual)
Action executed in operations
Outcome observed
Twin recommendation (NOT executed)
NOT EXECUTED
Logged for comparison
Decision Log: Human vs Twin
Human Decision & Outcome
Twin Recommendation & Logged Data
Compare outcomes + review differences
Where did they agree?
Where did they differ?
Which was correct?
A) Trust built with operators
B) Edge cases found → refine model/guardrails
C) Performance delta measured → evidence to scale
Production Readiness Gate: Twin consistently outperforms current practice?
YES
NO
Operationalize into workflows (alerts, work orders, approvals)
Refine + repeat shadow mode

The fifth and final step is operationalization: incorporating the twin's prescriptions into existing workflow processes. Operationalization does not necessarily involve complete automation, at least not initially. It could include presenting recommendations to the systems used by operators. It could include sending alerts to prompt human review. It could involve automatically creating work orders that humans would approve prior to execution. The primary objective is to ensure that the twin's output flows into the processes that are utilized in daily operations as opposed to merely residing on a dashboard waiting for someone to notice.

An Option-Based Approach

There's an important distinction between companies that achieve success and those that don't: treating your MVT pilots as options rather than commitments. A commitment represents a responsibility to take action whereas an option provides the ability to act (with the understanding that you may or may not).

In this way, investing a small amount of money in an MVT pilot allows you to determine if a concept has potential. Should the results prove positive, you can then invest additional resources. Conversely, if the results are poor, you have learned a lot about your idea and the associated costs were low.

Therefore, it's expected that many pilots will fail. However, it's also expected that you will kill these pilots quickly, so they're not lingering around, using up your company's resources, with uncertain outcomes. The worst-case scenario is not a failed pilot. It is a pilot stuck in limbo. A failed pilot costs money once and teaches a clear lesson. A zombie pilot bleeds resources indefinitely while blocking alternative investments, consuming executive attention, and generating neither proof nor disproof. Kill decisions are as important as scale decisions.

Set clear criteria prior to starting your pilot. Define the specific metrics you will be measuring during your pilot. Create a timeline for the pilot and once the information is collected, make a decision based upon the information. Grow the winners rapidly. Kill the losing pilots without hesitation.

Organizations that obtain the greatest value out of Digital Twins are often running multiple MVT pilots simultaneously. While the organization knows that not all pilots will be successful, they are confident that the winning pilots will generate enough revenue to offset the losses. Therefore, this approach creates a new risk/reward profile for the organization. Instead of placing the entire company's future at risk with one large initiative, the organization is taking several smaller risks with potentially high rewards. The potential loss for each of the initiatives is small while the potential gain is large.

However, even the most disciplined MVT process can be thwarted by two common pitfalls that warrant their own separate discussion. These include the tendency to allow data infrastructure projects to interfere with the delivery of immediate operational value and the trap of being bound by a particular vendor ecosystem before determining what you truly need. In the next section, we will discuss both issues and provide guidance on how to avoid them.

Chapter 9 – The Data Foundation

"Data is not a resource. It is infrastructure ignored until it collapses."□

The Many Plumbing Problems

You don't want to hear about pipes. But, you'll regret that choice. Think of your dream home: kitchen, view, space for entertaining. You never thought about pipe diameters or drainage grades. That said, once you mess up the plumbing, everything else becomes irrelevant. No matter how nice the kitchen is, no one will be able to use it if the sink won't drain.

That's the dirty little secret that Digital Twin vendors would prefer to ignore (the plumbing). The exciting part of the Digital Twin implementation (predictive algorithms, optimization engines, cool user interface) requires 20% of the overall efforts. The remaining 80% is plumbing: cleaning, tagging, integrating, transforming, validating. Plumbing refers to the effort to make sure information flows reliably from the source to the destination in a format that can be consumed by the receiving system. This is an unglamorous task and is essential to ensuring that the sophisticated models you build have enough good data to feed into the model.

Across many industries and implementations, data foundation work consistently consumes the majority of effort, often 60-80% of the total budget. Industry benchmarks from Gartner and Forrester corroborate what practitioners already know: the "boring" work of cleaning, tagging, and integrating data dwarfs the "exciting" work of building models. Most organizations grossly under-estimate the data foundation work, allocate inadequate time and money to it, and then have a major problem when the sophisticated models that they have built have nothing good to consume. The algorithms are ready to go. The data is not. Consequently, the project comes to a grinding halt while the engineers attempt to correct problems that should have been resolved months ago.

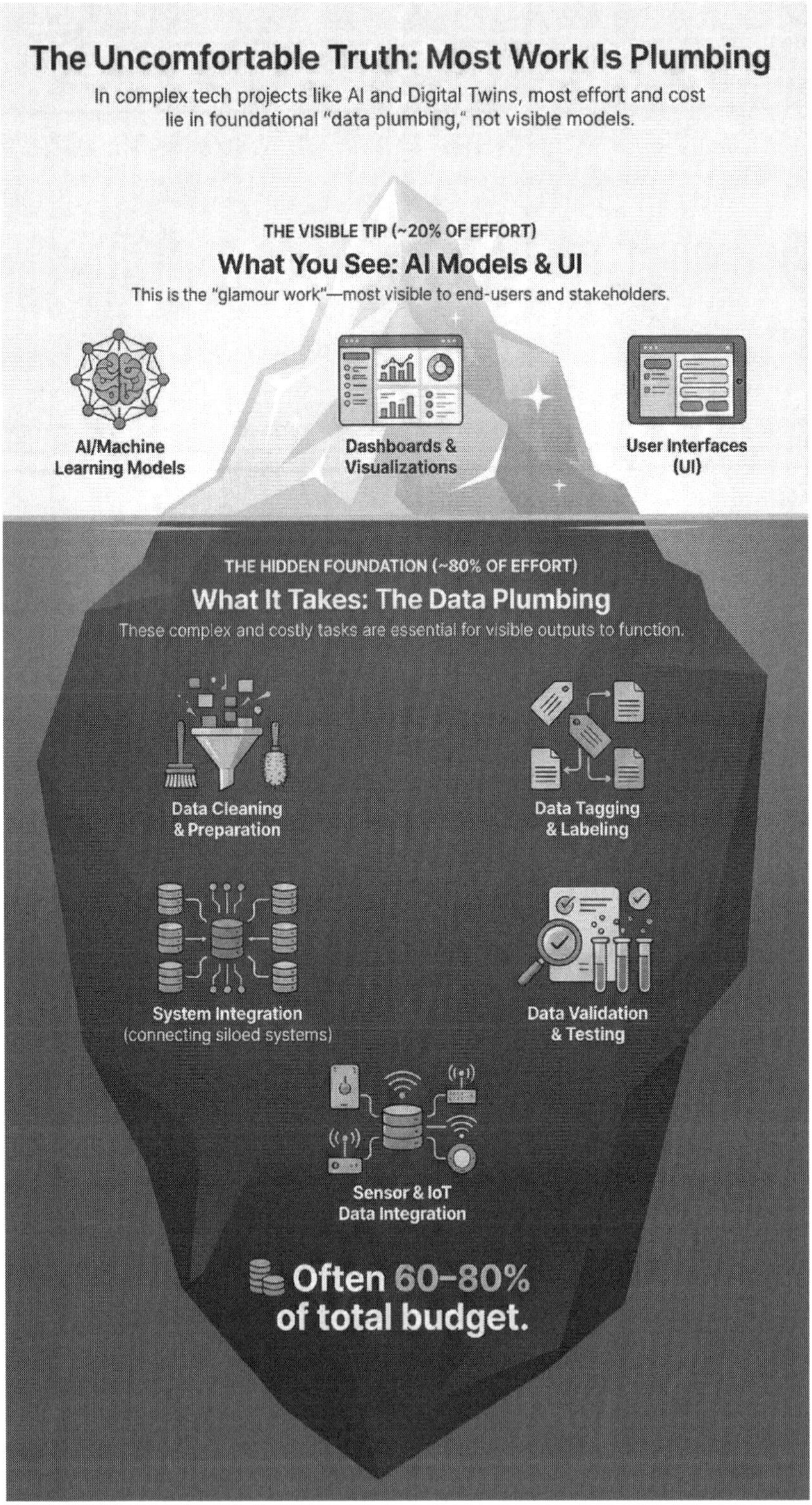
The Uncomfortable Truth: Most Work Is Plumbing
In complex tech projects like AI and Digital Twins, most effort and cost lie in foundational "data plumbing," not visible models.
THE VISIBLE TIP (~20% OF EFFORT)
What You See: AI Models & UI
This is the "glamour work"—most visible to end-users and stakeholders.
AI/Machine Learning Models
Dashboards & Visualizations
User Interfaces (UI)
THE HIDDEN FOUNDATION (~80% OF EFFORT)
What It Takes: The Data Plumbing
These complex and costly tasks are essential for visible outputs to function.
Data Cleaning & Preparation
Data Tagging & Labeling
System Integration (connecting siloed systems)
Data Validation & Testing
Sensor & IoT Data Integration
Often 60-80% of total budget.

The Allure of the Platform

In this complexity, there exists a vendor who makes a very attractive promise: "We've taken care of all that. Our platform handles data integration, model development, visualization and deployment. One contract. One Ecosystem. One Throat to Choke When Things Go Wrong." For many executives, who are already overwhelmed with the complexities associated with managing multiple technologies, this promise seems very attractive. Unfortunately, this path leads to a trap that has captured many organizations.

The way the trap works is that you choose a platform early, without fully understanding what your requirements are. The sales team has been very helpful. The demos have been very impressive. The references have been very reassuring. You enter into a large contract and begin to implement the platform. As you get further into the implementation process, you discover that the platform does well at some aspects of your business, but poorly at others. The features that seemed so great in the demos don't fit your specific workflow as well as you had hoped. The integrations that were touted as being easy-to-do are much harder to do than you had anticipated. At this point, however, you are invested in the platform. Switching would require throwing away months of work and tens-of-millions of dollars in sunk costs. Thus, you adjust your workflow to accommodate the platform, rather than vice versa.

Two years after selecting the platform, you are tightly coupled to it. Your data is in a format that is owned by the vendor. Your team is knowledgeable in the tools provided by the vendor. Your workflow is based upon the assumptions of the vendor regarding how work should flow. When a better tool becomes available, or when your business needs exceed what the platform can provide, the cost of changing to a new platform is prohibitively expensive. You are not a customer of the vendor anymore; you are captive.

The remedy is straightforward, yet difficult to execute: start with the problem you are trying to solve with the Digital Twin, and not the technology. Identify the decision you are trying to improve. Document the data needed to improve that decision. Then determine which technologies best support that need. Finally, when you do select technologies, ensure they are based upon open standards so that you may easily pivot as needed.

The Importance of Open Standards

Open standards play a strategic role in preventing lock-in. For twins with a 3D visualization layer, OpenUSD (Universal Scene Description) provides a widely adopted interchange format, originally developed by Pixar, now backed by a broad industry alliance. OpenUSD handles geometric representation and scene composition well. For operational semantics, i.e., the data models that describe process behavior, constraints, and decision logic, different standards apply: OPC-UA for industrial automation, Asset Administration Shell (AAS) for Industry 4.0 interoperability, FHIR for healthcare data exchange. The principle is consistent: own your data in formats that do not depend on a single vendor. With your data stored in open and documented formats, you are not dependent on a particular vendor to access it. With your models capable of exporting to standard formats, you can move them between platforms as your needs evolve.

While interoperability certainly prevents lock-in, it also enables composition. The best solution to your problem may consist of components from multiple vendors, each excellent at their respective functions, and composed using standard interfaces. Sensor networks from one vendor. Analytics engines from another. Visualization layers from a third. If each component communicates in a common language, you can create the best-of-breed solution rather than accept the compromises imposed by a single-platform solution.

Practically speaking for executives: when evaluating vendors, ask tough questions about data portability. Can you export your data in open formats? Can you run your models on different infrastructure? What happens to your intellectual property if you decide to leave? Vendors who struggle to answer these questions are signaling something about switching costs. If portability is weak, assume lock-in risk is high, and price that risk into your decision.

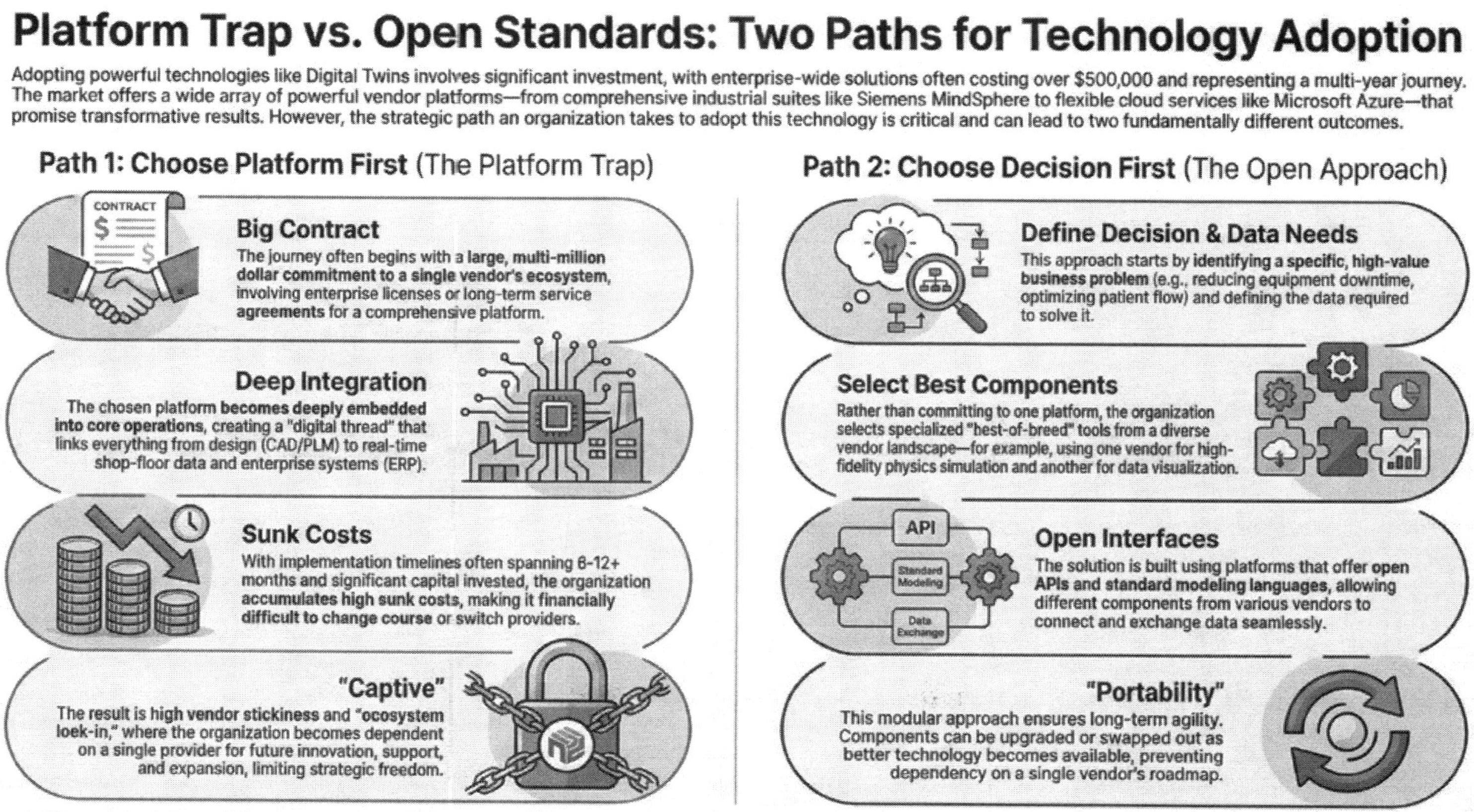
Platform Trap vs. Open Standards: Two Paths for Technology Adoption
Adopting powerful technologies like Digital Twins involves significant investment, with enterprise-wide solutions often costing over $500,000 and representing a multi-year journey. The market offers a wide array of powerful vendor platforms—from comprehensive industrial suites like Siemens MindSphere to flexible cloud services like Microsoft Azure—that promise transformative results. However, the strategic path an organization takes to adopt this technology is critical and can lead to two fundamentally different outcomes.
Path 1: Choose Platform First (The Platform Trap)
CONTRACT
Big Contract
The journey often begins with a large, multi-million dollar commitment to a single vendor's ecosystem, involving enterprise licenses or long-term service agreements for a comprehensive platform.
Deep Integration
The chosen platform becomes deeply embedded into core operations, creating a "digital thread" that links everything from design (CAD/PLM) to real-time shop-floor data and enterprise systems (ERP).
Sunk Costs
With implementation timelines often spanning 6-12+ months and significant capital invested, the organization accumulates high sunk costs, making it financially difficult to change course or switch providers.
"Captive"
The result is high vendor stickiness and "ocosystem lock-in," where the organization becomes dependent on a single provider for future innovation, support, and expansion, limiting strategic freedom.
Path 2: Choose Decision First (The Open Approach)
Define Decision & Data Needs
This approach starts by identifying a specific, high-value business problem (e.g., reducing equipment downtime, optimizing patient flow) and defining the data required to solve it.
Select Best Components
Rather than committing to one platform, the organization selects specialized "best-of-breed" tools from a diverse vendor landscape—for example, using one vendor for high-fidelity physics simulation and another for data visualization.
API
Standard Modeling
Data Exchange
Open Interfaces
The solution is built using platforms that offer open APIs and standard modeling languages, allowing different components from various vendors to connect and exchange data seamlessly.
"Portability"
This modular approach ensures long-term agility. Components can be upgraded or swapped out as better technology becomes available, preventing dependency on a single vendor's roadmap.

Data Challenges

While we've discussed several challenges in data, there is one other challenge related to data that needs some attention; this one is somewhat counter-intuitive: at times the data needed may not exist yet. Predictive models use historical trends to learn, however, what if you were attempting to forecast an event that occurs infrequently? An equipment failure that occurs every 5 years. A safety incident that has never occurred at your plant but has occurred at another location. An extreme condition that has never occurred at your plant but could occur in the future?

You cannot train a predictive model with data that does not exist. To put it differently, you could not until recently. With the emergence of Generative AI and specifically GANs (Generative Adversarial Networks), a new opportunity has emerged: synthetic data. Synthetic data can be generated through generative AI systems by learning the underlying patterns within available data and then generating plausible examples of events that have never actually occurred.

Synthetic data can be used to train a Digital Twin that identifies potential indicators of failure that it has never seen before. In addition, the Digital Twin can simulate thousands of extreme scenarios to prepare for those same extreme scenarios. Synthetic data does not replace operational data, but it expands what twins can learn from. The caveat is validation: synthetic failure modes must be tested against domain expertise and, where possible, against held-out real events. Without validation, you risk training models on artifacts of the generation process rather than patterns that reflect physical reality. Organizations leading the charge for Digital Twin capability are increasingly using synthetic data to develop predictive models for events they hope never occur.

However, one more challenge remains unaddressed: governance. As twins become more autonomous and capable, how do you ensure that twins operate in a reliable manner? How do you prevent a probabilistic model from producing deterministic errors? These questions take us to the fundamental difference between experimental pilot projects and production systems.

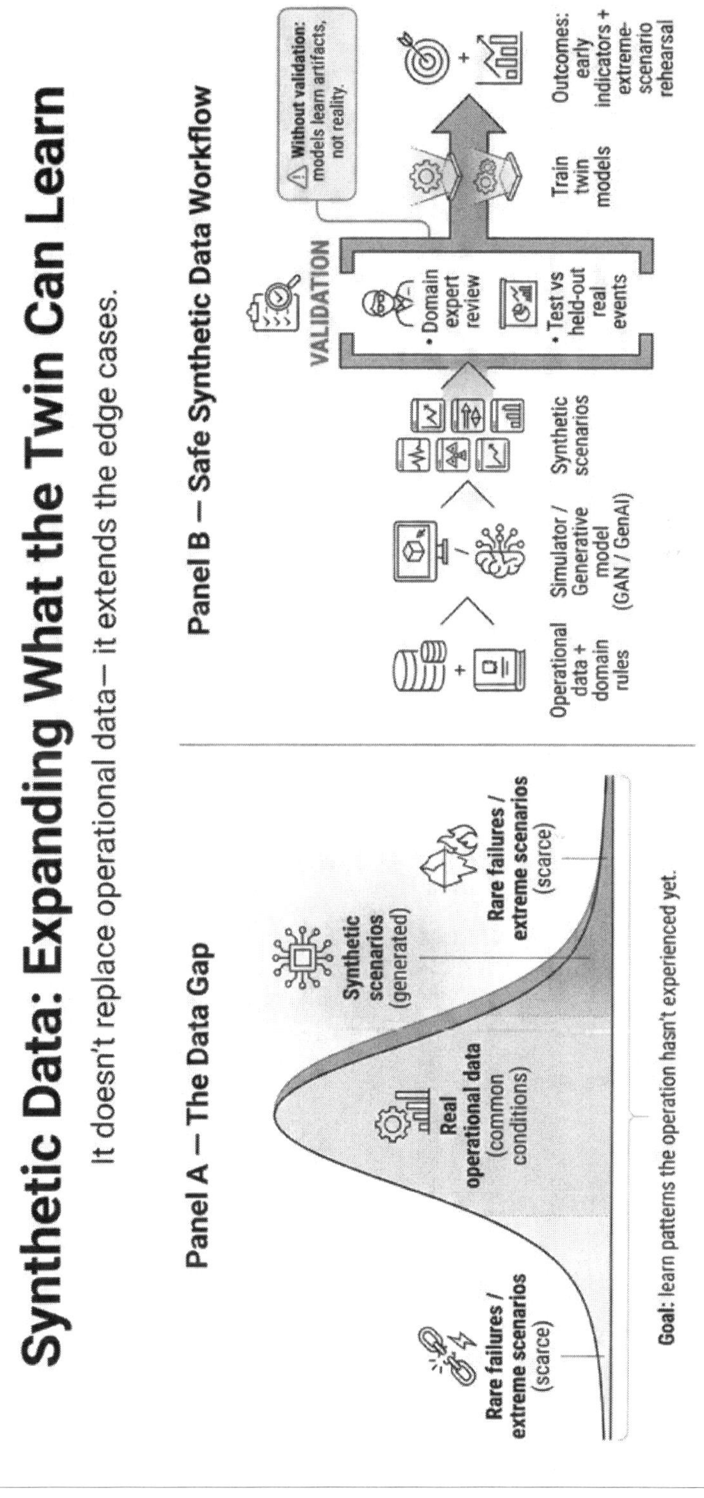
Synthetic Data: Expanding What the Twin Can Learn
It doesn't replace operational data— it extends the edge cases.
Panel A — The Data Gap
Rare failures / extreme scenarios (scarce)
Real operational data (common conditions)
Synthetic scenarios (generated)
Rare failures / extreme scenarios (scarce)
Goal: learn patterns the operation hasn't experienced yet.
Panel B — Safe Synthetic Data Workflow
Operational data + domain rules
Simulator / Generative model (GAN / GenAI)
Synthetic scenarios
VALIDATION
• Domain expert review
• Test vs held-out real events
Without validation: models learn artifacts, not reality.
Train twin models
Outcomes: early indicators + extreme-scenario rehearsal

Chapter 10 – Governance for Semi-Autonomy

"Autonomy is earned. Guardrails are how it is earned."□

Operational twins often include probabilistic components (e.g., failure predictions, demand forecasts, optimization under uncertainty) because the physical world is inherently uncertain. Deterministic physics simulations provide boundaries; probabilistic inference operates within them. Neither alone is sufficient. For example, when a Digital Twin says that there is a high chance (probability) of a pump failing in seventy-two hours, it is stating that there was a strong probability of that happening due to the patterns seen in the data. Probability does not equal certainty. Sometimes the pump could fail in forty-eight hours. Other times, it could continue to run for a month without failing. And other times, the model could simply be incorrect in ways that are expensive, embarrassing, or even dangerous.

The fear of what I refer to as "Digital Hallucinations" is also real. A machine learning model can develop outputs that are confidently incorrect. Predictions may appear logical, yet have absolutely nothing to do with how the physical world works. Models can extrapolate out from the training data into areas where the learned patterns no longer apply. Models can also identify spurious relationships between variables found in the past which have no basis in causation. If left unchecked, these hallucinations can cause companies to spend money, waste equipment, or put people in danger.

The Guardrail

The first line of defense against digital hallucinations is physics: hard constraints that prevent impossible outputs. But physics alone is not sufficient. A complete governance model layers physics constraints with validation tests, drift monitoring, escalation rules, and auditability. Each layer catches failures the others miss.

While algorithms may rely on statistical conveniences, the physical world operates under strict laws. Heat cannot spontaneously move from a cooler object to a warmer object. Mass cannot be created or destroyed within a closed system. Pressure cannot exceed material limits without consequence. Energy must be conserved. These constraints are absolute in a way that learned patterns never can be.

Physics-informed machine learning incorporates these physical laws directly into the model architecture. Thus, instead of permitting an algorithm to learn any pattern that fits the data, it is permitted to learn only those patterns that fit the physical laws. For example, the model may predict that a piece of equipment will fail, however it cannot predict that the failure violates any of the laws of thermodynamics. It may recommend an ideal temperature for the equipment to run at, but it cannot suggest a setting that would require negative energy consumption.

This hybrid approach combines the strengths of both paradigms. Physics provides the guardrails that prevent impossible predictions. Machine learning captures the subtle, nonlinear patterns that physics alone might miss. The result is a model that is both flexible enough to learn from data and constrained enough to remain grounded in reality. When executives ask what prevents hallucinations, this is the first answer: we've built the laws of physics into the system's DNA.

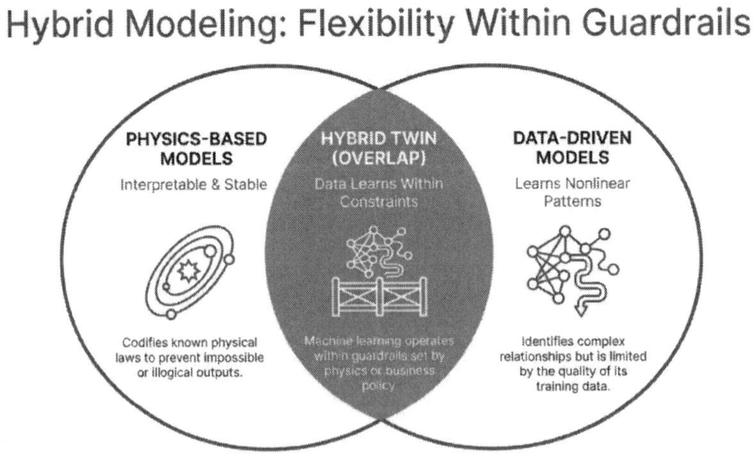

Physics constraints apply cleanly to industrial twins such as pumps, turbines, reactors, logistics networks governed by thermodynamics and material limits. But what about patient flow twins? Customer pricing twins? Revenue optimization systems? These operate under different constraints: regulatory rules, contractual commitments, resource limits, business logic. The governance principle remains the same: Embed hard boundaries that prevent impossible or prohibited outputs, but the boundaries are policy and law rather than physics. A hospital twin cannot recommend staffing below mandated nurse-to-patient ratios. A pricing twin cannot propose terms that violate contract law. The discipline is identical; the constraint vocabulary differs.

Governance for Semi-Autonomy

Defense-in-depth: each layer catches failures the others miss.

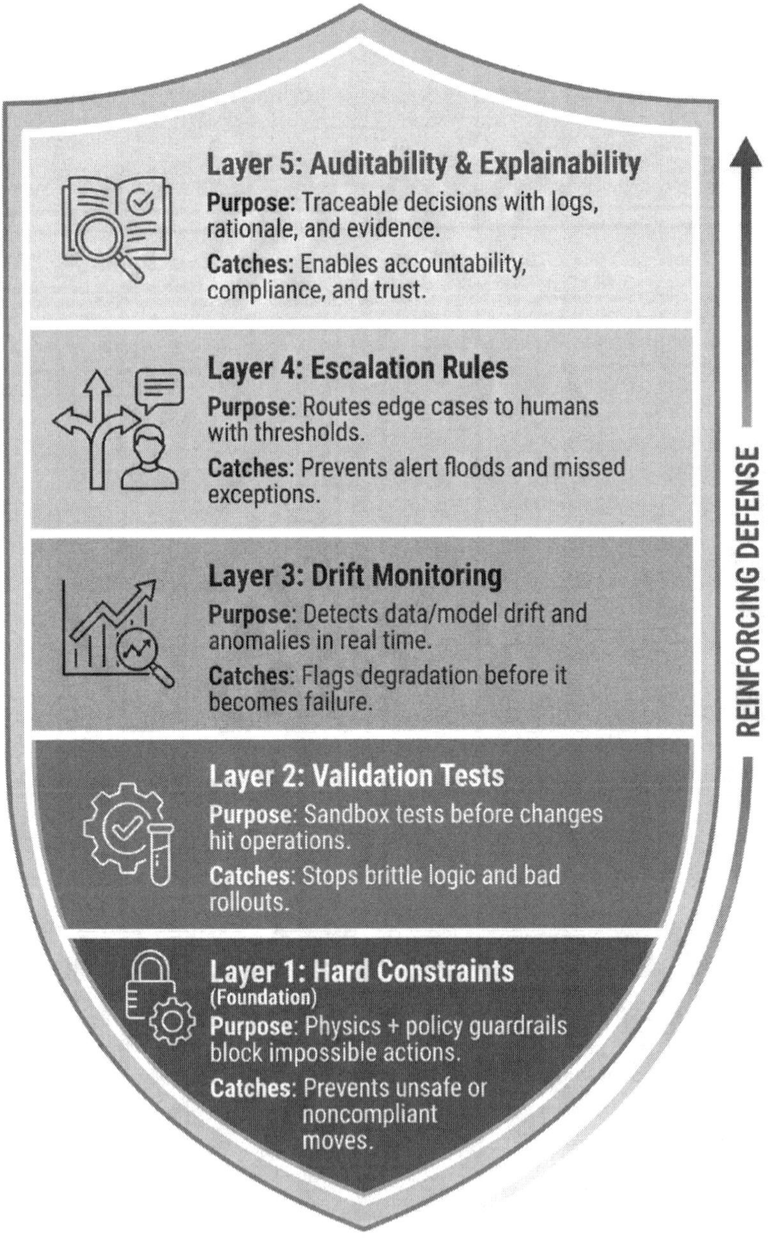

The Autonomy Charter

Physics limits what the twin may know, however you also must limit what the twin can do. This is the purpose of the Autonomy Charter, a one-page governance document that defines the bounds of the systems' authority. With clarity, the charter answers three fundamental questions:

- What can the system independently execute?
- What does the system need to recommend to a human for approval?
- Under what circumstances must the system immediately escalate?

A good charter cannot be a technical document. It needs to be written in terms that business leaders can understand and accept, since those individuals are responsible for what the system does. A well-crafted charter may specify that the twin can autonomously adjust production settings within a specified range, while requiring human approval for adjustments outside that range. It may provide for autonomous creation of work orders for routine maintenance but require a person to approve actions affecting safety-critical equipment. It may outline specific escalation criteria (e.g., confidence level, cost threshold, novelty of the situation).

The primary purpose of the charter is to establish the boundaries explicitly and intentionally, as opposed to implicitly and by accident. Every company transitioning toward semi-autonomous systems will eventually find itself in a situation where the system has done something that was unanticipated. When that occurs, the desired response to the question of "why did it do that?" is to respond with "because we authorized it to," as opposed to "we were unaware it could."

Autonomy Charter

Define what the system may execute, what it must recommend, and when it must escalate.

Context

Twin/System Name: ....

Decision Scope (one sentence): ....

Boundary (In / Out): ....

Decision Rights

A

EXECUTE

System acts independently (within guardrails)

Allowed Actions:

....

....

....

Limits / Ranges:

....

Examples:

....

....

B

RECOMMEND

System proposes; human approves

Recommended Actions:

....

....

....

Approval Role:

....

Examples:

....

....

C

ESCALATE

System routes immediately to an owner

Escalation Triggers:

....

....

....

Escalation Owner:

....

Examples:

....

....

Thresholds & Triggers (Defaults)

- ☐ Confidence below ___% → Escalate
- ☐ Estimated impact above $_____ → Escalate
- ☐ Novel / out-of-distribution scenario → Escalate
- ☐ Safety-critical equipment / compliance flag → Escalate

Accountability

Business Owner: ....

Audit / Log Requirement: ....

Effective Date: .... **Review Cadence:** ....

The Explainability Requirement

There is one more governance element that distinguishes production-ready twins from experimental prototypes: auditability. The system must be capable of providing an explanation for why it made a particular recommendation. Not just what it recommended, but the rationale behind that recommendation.

- Which inputs had the greatest influence?
- What patterns resulted in the predictive result?
- How confident is the model, and what would impact the confidence in the model?

Auditability fulfills multiple functions. For operators, auditability establishes trust. A maintenance technician is more likely to take action on a recommendation if he/she/they can visually observe that the recommendation is predicated upon recognizable vibration patterns, and not simply a black-box prediction. For managers, auditability facilitates oversight. Should anything go awry, the decision-making process can be traced back to its inputs, and the systemic errors that lead to the incorrect decision can be identified. For regulators and auditors, auditability provides accountability. In industries where decisions are subject to review, the capability to recreate and substantiate the reasoning process is essential.

From a practical standpoint, it is prudent to be skeptical of systems that cannot explain how they arrived at a conclusion. If a vendor advises that their model is so complex that it cannot be interpreted, that is not a desirable attribute; it is a liability. Today's most advanced twins being used in production environments are constructed from the ground-up to be interrogated; to expose the logic employed in reaching a determination; to illustrate the methodology as if they were students completing a test.

The Explainability Requirement

A production twin must show its work: not just WHAT, but WHY.

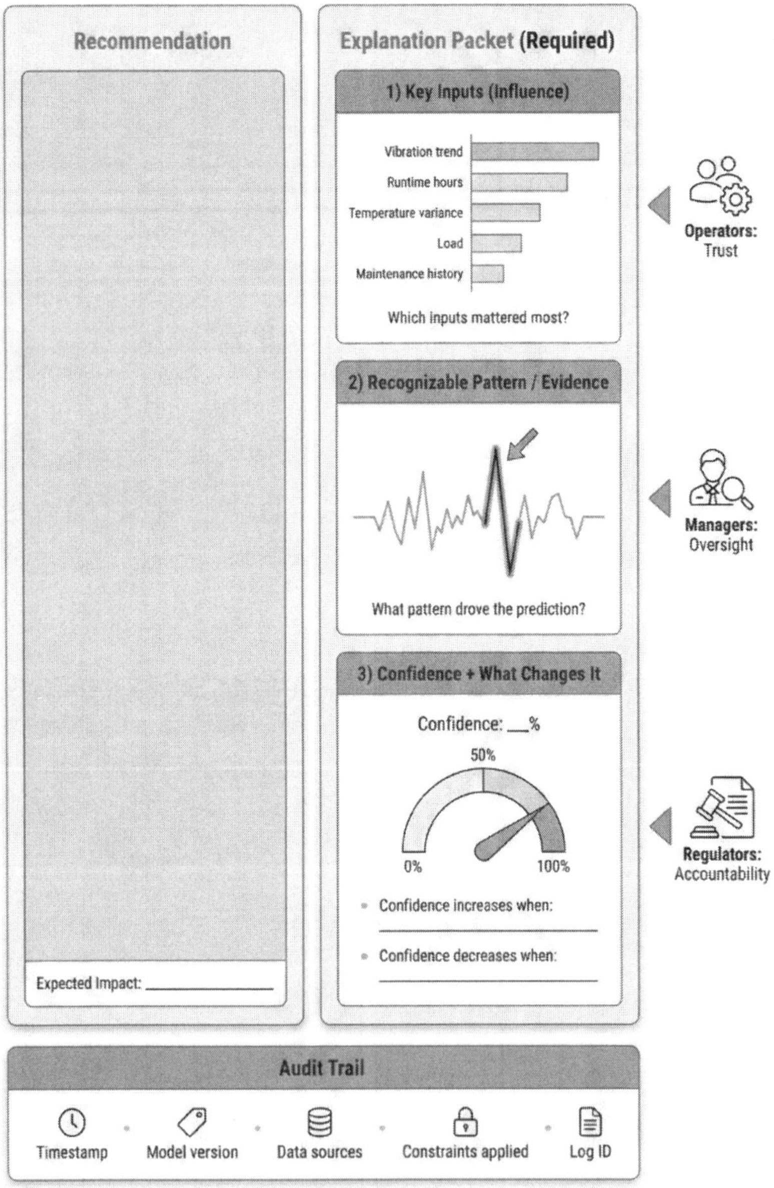

With the proper governance in place (i.e., physics-informed models, clearly defined autonomy boundaries, and explainable reasoning), you have the foundation to create a twin that can be relied upon to assume genuine operational responsibility. However, having knowledge of the principles does not equate to execution. In the final chapter, all concepts will be incorporated into useful tools: a readiness scorecard to gauge your current status, a canvas to scope your initial twin, and a forward-thinking perspective on where this technology is headed next.

Chapter 11 – The CEO's Toolkit and the Future Outlook

"The CEO's job is not to choose a model. It is to choose the decision the model will own."□

We've covered a lot of ground. The conceptual framework. The architecture. The implementation playbook. The governance principles. Now it's time to translate all of that into tools you can actually use. Because frameworks without application are just intellectual decoration, and the goal of this book has never been to make you smarter about Digital Twins. It's been to help you build one that works.

The Readiness Scorecard

Before investing resources into a Digital Twin initiative, you need an honest assessment of where your organization currently sits relative to where you want to sit. Not where you believe you should be or where your strategic plan states you should be, but rather where you are today. I adopt a 10-question scorecard, which measures 3 areas: Data Readiness, Operational Clarity and Cultural Receptiveness.

Data Readiness

In the area of Data Readiness, you need to determine whether the signals needed to support your targeted decision are being collected. Not whether they could be collected with sufficient funding, but whether they are currently being captured. You need to evaluate your data quality honestly: how much "clean-up" would be necessary before those signals could feed into a model? Lastly, you need to establish an understanding of the latency associated with the time it takes for events in the physical world to appear in your data systems.

Digital Twin Readiness Scorecard

A self-assessment tool for businesses to evaluate their preparedness across key areas before implementing a digital twin solution.

DATA READINESS

- ○ **Key Signals Exist**
  You can identify and access the critical data streams from your physical assets/processes.
- ○ **Data Quality is Sufficient**
  Data is accurate, complete, and reliable enough to build a trustworthy model.
- ○ **Latency is Acceptable**
  Data can be collected and processed quickly enough for the intended application.

OPERATIONAL CLARITY

- ○ **Clear Decision to Support**
  You know the specific operational decision the digital twin will improve or automate.
- ○ **Constraints are Defined**
  Physical, operational, and financial limitations of the system are well understood.
- ○ **Baseline Metrics Exist**
  You have current performance metrics to measure against for proving value.

CULTURAL RECEPTIVENESS

- ○ **Operator Trust**
  Front-line employees are willing to trust and act on the twin's recommendations.
- ○ **Manager Acceptance**
  Middle management understands the value and supports the necessary process changes.
- ○ **Executive Sponsorship**
  There is strong, visible support and funding from senior leadership.

Operational Clarity

In the area of Operational Clarity, you need to establish clarity around the decision you are trying to Improve. Specifically, can you clearly articulate the exact decision you are trying to improve, who makes the decision today and how would you measure improvement in that decision? In addition, do you have a solid understanding of the constraints that limit that decision, the rules that must be followed regardless of what recommendations the model makes? Finally, do you have a baseline established by which you can measure the performance of your Twin?

Cultural Receptiveness

In the area of Cultural Receptiveness, you need to establish whether your organization is prepared to act differently based on algorithmic recommendations. Will operators trust a machine's judgment? Will managers accept that a model may perform better than their intuition? Does your organization have sufficient executive sponsorship to overcome the resistance that will inevitably arise? Your culture doesn't change simply because a project launches; projects fail because culture doesn't change.

Organizations that perform well in all three areas can rapidly transition into implementation mode. Organizations that have gaps in one area can continue to implement while simultaneously addressing the gap in the other area. Organizations that have gaps in all three areas should likely invest in foundation building prior to attempting to create a twin. There is no shame in such an assessment; it is significantly better to know now versus discovering it after spending millions. Honest self-assessment is the first step towards wisdom and rushing into a twin without establishing readiness is the quickest route to becoming part of the large number of failed AI initiatives.

The Opportunity Canvas

After establishing that your organization is ready, you need a structured way to scope your Minimum Viable Twin (MVT). I use a tool that I refer to as the Twin Opportunity Canvas: a single-page template that requires you to clarify the fundamental components of your twin prior to any coding commences.

The Canvas starts with the boundary. What's inside the system you're modeling and what's outside? Where does your twin's jurisdiction end and the external environment begin? Getting this wrong at the start creates confusion that compounds through the entire implementation.

Next is the Decision. Articulate the decision in a single sentence. Not "optimize maintenance," but "determine when to service pump #7." Not "improve patient flow," but "identify which patients are ready for discharge this shift." If you cannot articulate the decision in a single clear sentence, you have not sufficiently defined the Scope of your Twin.

After that comes the Value Hypothesis. What outcome will improve if this decision gets better? By how much? How will you measure it? This forces you to link the technical work to business impact prior to starting, not after. A twin without a clear value hypothesis is merely a science project, not a business initiative.

While presented sequentially, these elements inform each other. Defining constraints may force you to redraw the boundary. Articulating the value hypothesis may reveal that you have scoped the wrong decision. Expect to iterate. The Canvas is not a one-pass exercise; it is a forcing function for clarity that typically requires two or three revisions before the team achieves alignment. When completed honestly, the canvas serves as both a project charter and accountability document that maintains alignment among all parties involved as the implementation proceeds.

A World of Controlled Semi-Autonomy

The timeline is uncertain, but the direction is clear: More bounded operational decisions will be delegated to semi-autonomous systems under explicit guardrails. In the near term, this expansion will concentrate on high-frequency tasks with measurable outcomes, where systems can act within constraints and escalate exceptions to human owners. The executive question is not whether semi-autonomy arrives, but where it will be permitted first, what limits will govern it, and who will own results.

Twin Opportunity Canvas

This canvas is a strategic planning tool designed to capture the essential elements of a new opportunity or project on a single page. It helps teams establish clear boundaries, define success, and identify critical constraints before committing significant resources.

1. BOUNDARY

What is IN SCOPE?

List the specific activities, features, and components that this initiative will address.

____

____

____

What is OUT OF SCOPE?

Explicitly state what will not be addressed to prevent scope creep.

____

____

____

2. DECISION

State the core decision to be made.

Formulate a single, clear sentence that this work will help answer or validate.

____

____

3. VALUE HYPOTHESIS

Formulate the value hypothesis.

Use the format: We believe [doing this] for [these people] will achieve [this outcome].

We believe _______________ for _______________ will achieve _______________.

Purpose: To clearly define how the value hypothesis will be measured and validated.

Metric	Target	Measurement Method
e.g., Production Downtime	e.g., Reduce by 15%	e.g., Weekly system logs
________	________	________
________	________	________
________	________	________

4. CONSTRAINTS & GUARDRAILS

What are the Non-Negotiables?

List fixed constraints such as budget limits, deadlines, or regulatory requirements.

____

____

____

____

What are the Escalation Triggers?

Define conditions that would require immediate escalation to leadership (e.g., budget overrun >10%).

____

____

____

____

This shift is less about adopting new tools than about redesigning decision loops: defining constraints, embedding prescriptions into workflows, assigning decision rights, and measuring outcomes. Organizations that treat AI as an operating-model change are more likely to capture durable performance gains than those that treat it as another reporting upgrade.

There will be significant competitive advantages associated with the speed of decision-making. When your systems can react to changes in the environment much quicker than your competitor can identify those changes, speed is a structural advantage. When your systems can self-adapt to disruptions while your competitors' systems are unable to due to manual intervention, resilience is a structural advantage. Finally, consistency is a structural advantage when all decisions made thousands of times each day are optimized versus merely good enough (satisficed). The gap between leaders and laggards in operational AI will widen, not narrow, because the capabilities compound over time.

Those organizations who successfully digitize their operational DNA will win in the next decade. They'll build digital representations of their processes so complete that they can simulate the future before it happens, test decisions before committing to them, and learn from failures that never actually occurred. The organizations that develop these digital models will view the twin as a continuous evolving capability that will become more advanced and more autonomous over time as it gains more trust based upon actual results. The twin will begin to represent the way the organization thinks, and not simply another tool the organization uses.

The organizations that succeed will measure progress by decision quality and cycle time, not by technology adoption milestones. A twin is not a system you install; it is a capability you build, refine, and expand as trust accumulates. The competitive advantage compounds quietly: fewer surprises, faster responses, better use of assets. Organizations that focus solely on the technology will continue to struggle. Those that focus on the decision loop will win.

As we conclude this book, I would like to provide you with a basic decision-making framework regarding where to begin and a challenge regarding what to accomplish next.

Conclusion: The Quiet Advantage

"Advantage will compound where the distance from signal to response is shortest."

Digital Twins are not a marketing fad; nor are they another 'next big thing' for consultants to peddle or vendors to demonstrate. They represent a more enduring and practical mechanism for creating governed decision-making loops that connect data to action in a manner that dashboards have never been able to achieve. Organizations that are successfully leveraging Digital Twins are not chasing hype. They are using Digital Twins to solve very tangible and very important operational issues with quantifiable results, one decision at a time.

The benefit that these organizations are creating is quiet. In fact, it does not even generate much fanfare via the form of press releases or keynote presentations. Instead, the benefit is generated in the form of incremental improvements to processes, which ultimately result from thousands of decisions. For example, a maintenance activity is performed at exactly the right time, a patient is discharged as soon as they are medically fit, a shipment is routed around a potential delay before it becomes a real issue, a leak is identified and corrected before it becomes evident. While each of these individual improvements may appear minor, collectively they provide a fundamental paradigm shift in terms of how successful organizations operate, a way that competitors can see in the results but struggle to replicate because they don't understand the architecture beneath.

These organizations understand something that many of their competitors are still trying to grasp. The value of Artificial Intelligence (AI) is not based upon the quality of the model itself. It's based on the organization's ability to leverage its operating model to create a seamless connection between predictive analytics and actionable intelligence, between insight and execution, and between what is known and what is done. An AI model that generates reports that are ignored by stakeholders creates zero value. Conversely, a relatively simple model that generates the correct actions at the right moments creates enormous value. This is not about achieving high levels of sophistication within the model itself. It's about developing an architecture that focuses relentlessly on closing the loop.

So here is my challenge to you. Before you close this book, to take a few minutes to determine the two operational decisions that are consuming the largest amount of expense or generating the greatest level of risk in your organization. These are the operational decisions that are made numerous times throughout the course of a business day, typically by individuals who are unaware of the fact that they are making decisions. Examples of such operational decisions include:

- Scheduling related decisions
- Maintenance timing decisions
- Resource allocation decisions
- Routing determination decisions
- Triage-related decisions

Then ask yourself: what would it mean to make those decisions 5% better, consistently, every single time? Once you run the numbers, you will likely find that the financial impact of improving these decisions by just 5% will be significantly larger than you expected, since these decisions are repeated multiple times during the course of a business day, and the cumulative effect of these improvements is what drives significant economic changes to an organization.

This is where you should begin. Don't begin with a grand vision for transforming your organization or a multi-year road map. Instead, begin with two decisions that you understand well enough to improve. Use the Readiness Scorecard to evaluate your organization's current state and use the Opportunity Canvas to document the problem(s) you want to address. Create a Minimum Viable Digital Twin. Test it in Shadow Mode until you feel confident that the results are accurate. Once you are satisfied with the results, then operationalize and expand the use of the Digital Twin. Let your successes earn the right to continue growing.

The Digital Twin Mirror has been created. It is now time for you to look into it and visualize what your operations could potentially become. The enabling technologies are available today. The Digital Twin frameworks have been tested and validated. The only remaining question is: will you be part of the group of organizations that create the quiet advantages, or will you be looking back over the next ten years wondering how your competitors were able to gain such a significant competitive advantage?

References

Academic Journals, Books & Research Papers

- **Angel, James J., et al.** 2011. "Equity Trading in the 21st Century." *Quarterly Journal of Finance.*
- **Baron, Matthew, et al.** 2019. "Risk and Return in High-Frequency Trading." *Journal of Financial and Quantitative Analysis.*
- **Benos, Evangelos, and Satchit Sagade.** 2016. "Price Discovery and the Cross-Section of High-Frequency Trading." *Journal of Financial Markets.*
- **Biais, Bruno, and Paul Woolley.** 2012. "High Frequency Trading." *IDEI Working Papers.*
- **Billio, Monica, et al.** 2012. "Econometric Measures of Connectedness and Systemic Risk in the Finance and Insurance Sectors." *Journal of Financial Economics.*
- **Bisias, Dimitrios, et al.** 2012. "A Survey of Systemic Risk Analytics." *Annual Review of Financial Economics.*
- **Boschert, Stefan, and Roland Rosen.** 2016. "Digital Twin—The Simulation Aspect." *Mechatronic Futures.*
- **Brogaard, Jonathan, et al.** 2014. "High-Frequency Trading and Price Discovery." *The Review of Financial Studies.*
- **Brogaard, Jonathan, et al.** 2015. "Trading Fast and Slow: Colocation and Liquidity." *The Review of Financial Studies.*
- **Budish, Eric, et al.** 2015. "The High-Frequency Trading Arms Race: Frequent Batch Auctions as a Market Design Response." *The Quarterly Journal of Economics.*
- **Chaboud, Alain P., et al.** 2014. "Rise of the Machines: Algorithmic Trading in the Foreign Exchange Market." *Journal of Finance.*
- **Crespi, Noel, et al. (Eds).** 2023. *The Digital Twin.* Springer.
- **Easely, David, et al.** 2012. "Flow Toxicity and Liquidity in a High-Frequency World." *The Review of Financial Studies.*
- **Gelernter, David.** 1991. *Mirror Worlds.* Oxford University Press.
- **Girling, Phillippa X.** 2013. *Operational Risk Management.* Wiley.
- **Glosten, Lawrence R., and Paul R. Milgrom.** 1985. "Bid, Ask and Transaction Prices in a Specialist Market with Heterogeneously Informed Traders." *Journal of Financial Economics.*
- **Grieves, Michael, and John Vickers.** 2017. "Digital Twin: Mitigating Unpredictable, Undesirable Emergent Behavior in Complex Systems." *Transdisciplinary Perspectives on Complex Systems.*
- **Harris, Larry.** 2003. *Trading and Exchanges: Market Microstructure for Practitioners.* Oxford University Press.
- **Hasbrouck, Joel, and Gideon Saar.** 2013. "Low-Latency Trading." *Journal of Financial Markets.*
- **Hendershott, Terrence, et al.** 2011. "Does Algorithmic Trading Improve Liquidity?" *The Journal of Finance.*

- **Jarrow, Robert A., and Philip Protter.** 2012. "A Dysfunctional Role of High Frequency Trading in Electronic Markets." *International Journal of Theoretical and Applied Finance.*
- **Jones, Charles M.** 2013. "What Do We Know About High-Frequency Trading?" Columbia Business School.
- **Lewis, Michael.** 2015. *Flash Boys.* W. W. Norton & Company.
- **Liu, Mangnan et al.** 2021. "Review of Digital Twin About Concepts, Technologies, and Industrial Applications." *Journal of Manufacturing Systems.*
- **Mahajan, Vijay.** 2010. "Innovation Diffusion." *Wiley International Encyclopedia of Marketing.*
- **Meech, John, and Juliana Parreira.** 2011. "An Interactive Simulation Model of Human Drivers to Study Autonomous Haulage Trucks." *Procedia Computer Science.*
- **Semeraro, Concetta et al.** 2021. "Digital Twin Paradigm: A Systematic Literature Review." *Computers in Industry.*
- **VanDerHorn, Eric, and Sankaran Mahadevan.** 2021. "Digital Twin: Generalization, Characterization and Implementation." *Decision Support Systems.*

Government, Regulatory & Standards Bodies

- **Consultative Committee for Space Data Systems (CCSDS).** 2019. *Report Concerning Space Data System Standards.* CCSDS 350.7-G-2.
- **Department of Energy (US).** 2016. "Distribution Automation: Results from the Smart Grid Investment Grant Program." *Smart Grid Investment Grant Program.*
- **European Central Bank.** De Bandt, Olivier, and Philipp Hartmann. 2000. "Systemic Risk: A Survey." *ECB Working Paper No. 35.*
- **Federal Aviation Administration (FAA).** 2020. "GE9X Type Certificate Data Sheet E00095EN." *Dynamic Regulatory System.*
- **INCOSE – International Council on Systems Engineering.** 2021. "Systems Security Engineering: Mission and Objectives." *INCOSE Working Group/Web Resource.*
- **International Organization of Securities Commissions (IOSCO).** 2011. "Regulatory Issues Raised by the Impact of Technological Changes on Market Integrity and Efficiency." *IOSCO Technical Committee Report*
- **NASA.** 2019. *Space System Protection Standard.* NASA-STD-1006.
- **NASA.** Allen, B. Danette. 2021. "Digital Twins and Living Models at NASA."
- **National Institute of Standards and Technology (NIST).** Ross, R., et al. 2022. "Systems Security Engineering: Engineering Trustworthy Secure Systems." *NIST SP 800-160 Vol. 1, Rev. 1.*
- **National Institute of Standards and Technology (NIST).** Ross, R., et al. 2021. "Developing Cyber-Resilient Systems: A Systems Security Engineering Approach." *NIST SP 800-160 Vol. 2, Rev. 1.*
- **National Institute of Standards and Technology (NIST).** David, I., et al. 2024. "Interoperability of Digital Twins: Challenges, Success Factors, and Future Research Directions." *NIST Technical Publication.*

- **NREL (National Renewable Energy Laboratory).** Pratt, A., et al. 2021. "Defining a Use Case for the ADMS Test Bed: Fault Location, Isolation, and Service Restoration with Distributed Energy Resources." *NREL Conference Paper.*
- **U.S. Department of Justice.** 2015. "Futures Trader Charged With Illegally Manipulating Stock Market, Contributing to the May 2010 Market 'Flash Crash'." *Official Press Release.*
- **U.S. Securities and Exchange Commission (SEC).** 2010. "Findings Regarding the Market Events of May 6, 2010."
- **U.S. Securities and Exchange Commission (SEC).** Gerig, Austin. 2015. "High-Frequency Trading Synchronizes Prices in Financial Markets." *SEC DERA Working Paper.*

Industry Reports & Whitepapers

- **Accenture.** 2025. *The Digital Twin Revolution and its M&A Momentum.*
- **Advisory Board.** "Johns Hopkins Created 16 Beds' Worth of Capacity Without Adding a Single Bed."
- **Aerospace Corporation, The.** 2021. *Defending Spacecraft in the Cyber Domain.*
- **Deloitte.** "From Manufacturing to Medicine: How Digital Twins Can Unlock New Industry Advantages."
- **Deutsche Bank Research.** 2016. *Research Briefing: High-Frequency Trading.*
- **Bojinov, Iavor I., Karim R. Lakhani, and David Lane.** 2025. "JPMorganChase: Leadership in the Age of GenAI." *Harvard Business School Case.*
- **Manufacturing Leadership Council.** 2020. "The Transformative Power of Plant Digital Twins." *Manufacturing Leadership Journal.*
- **McKinsey & Company.** 2025. "An Inside Look at How McKinsey Helped DBS Become an AI-Powered Bank." *McKinsey Case Study.*
- **McKinsey & Company.** 2024. "Digital Twins and Generative AI: A Powerful Pairing." *McKinsey Digital.*
- **McKinsey & Company.** 2023. "Digital Twins: The Art of the Possible in Product Development and Beyond." *McKinsey Digital.*
- **MITRE Corporation.** 2020. Visner, S., and S. C. Kordella. *Cyber Best Practices for Small Satellites.*
- **Siemens AG.** 2020. *Executing Digital Twins: How to Realize the Vision.*

Corporate Case Studies

- **Amazon Web Services (AWS).** 2024. "Using Simulation and Digital Twins to Increase Productivity and Quality and Reduce Cost."
- **Chevron.** 2012. "Chevron's CIO Talks Transformation." *CIO.com.*
- **GE HealthCare.** 2025. "The Queen's Health Systems Saves $20 Million in the First Year After Opening a Command Center."
- **GE Research.** 2018. "GE Global Research Awarded $12.6 Million from US DOE to Develop Quantum-level Cybersecurity."

- **General Motors.** 2025. "GM Builds on U.S. Manufacturing Investments with $250M Commitment in Parma Metal Center."
- **Mayo Clinic Platform.** 2024. "Digital Twin Technology Has Potential to Redefine Patient Care."
- **NVIDIA.** 2021. "Using Hybrid Physics-Informed Neural Networks for Digital Twins in Prognosis and Health Management." *NVIDIA Technical Blog.*
- **Philips.** 2025. "The Rise of the Digital Twin: How Healthcare Can Benefit."
- **Rio Tinto.** 2019. "Mine of the Future." *Rio Tinto Official Report.*
- **Schneider Electric / AVEVA.** 2025. "Schneider Electric, AVEVA and ETAP Join Alliance for OpenUSD to Advance Digital Twins."
- **Tampa General Hospital.** 2020. "Tampa General Hospital and GE Healthcare's CareComm Saves $40 Million."

News & Media

- **Bloomberg.** 2023. "NYSE Computer Error Prompts Cancellation of Almost 20,000 Trades."
- **CIO.com.** 2012. Gallant, J. "Chevron's CIO Talks Transformation and Why IT Leaders Should Smile."
- **CNBC.** 2013. Clinch, Matt. "Italy Launch Tax on High-Frequency Transactions."
- **Economist, The.** 2016. "Siemens and General Electric Gear Up for the Internet of Things."
- **Financial Times.** 2013. Stafford, P. "Italy Introduce Tax on High-Speed Trade and Equity Derivatives."
- **Wall Street Journal.** 2012. Strasburg, J., and Bunge, J. "Loss Swamps Trading Firm." (Knight Capital Case)

Made in the USA
Monee, IL
01 May 2026

49351279R00066